Sky Alert!

When Satellites Fail

Les Johnson

Sky Alert!

When Satellites Fail

 Springer

Published in association with
Praxis Publishing
Chichester, UK

Les Johnson
Madison
Alabama
USA

SPRINGER–PRAXIS BOOKS IN POPULAR SCIENCE
SUBJECT *ADVISORY EDITOR*: Stephen Webb, B.Sc., Ph.D., M.Inst.P., C.Phys.

ISBN 978-1-4614-1829-0   ISBN 978-1-4614-1830-6 (eBook)
DOI 10.1007/978-1-4614-1830-6
Springer New York Heidelberg Dordrecht London

Library of Congress Control Number: 2012947008

Cover design: Jim Wilkie
Project copy editor: Christine Cressy
Typesetting: BookEns, Royston, Herts., UK

Printed on acid-free paper

Springer is part of Springer Science+Business Media (www.springer.com)

Dedicated to my parents, Charles and June Johnson.
Although they are no longer on this Earth,
they shall never be forgotten.

Acknowledgment

I would like to thank my long-time friend and colleague, Dr James Woosley, for contributing Chapter 15 and for being part of the brainstorming sessions that ultimately led to this book being published.

About the Author

Les Johnson is an author of popular books about space exploration and a science-fiction novelist. He also serves as the Deputy Manager for NASA's Advanced Concepts Office at the Marshall Space Flight Center in Huntsville, Alabama, where he was a co-investigator for the JAXA T-Rex Space Tether Experiment and PI of NASA's ProSEDS Experiment.

His popular science books include *Living Off the Land in Space* (Springer/Copernicus, 2007), the 2008 PROSE Award finalist, *Solar Sails: A Novel Approach to Interplanetary Travel* (Springer/Copernicus, 2008), and *Paradise Regained: The Regreening of Earth* (Springer/Copernicus, 2009). His science-fiction novel, *Back to the Moon*, was published in hardcover (2010) and paperback (2012) by Baen. His latest science-fiction book, *Going Interstellar*, was published in May 2012.

During his career at NASA, he served as the Manager for the Space Science Programs and Projects Office, the In-Space Propulsion Technology Program, and the Interstellar Propulsion Research Project. He twice received NASA's Exceptional Achievement Medal and has three patents.

He earned his master's degree in physics from Vanderbilt University (Nashville, TN) and his BA at Transylvania University (Lexington, KY). He participated in the International Space University's 1993 Summer Session Program in Toulouse, France. Les has numerous peer-reviewed publications and was published in *Analog*. He is a frequent contributor to the *Journal of the British Interplanetary Society* and has been a member of the American Institute of Aeronautics and Astronautics, National Space Society, the World Future Society, and MENSA.

Contents

Chapter Summaries

Much of the world now depends upon space technology economically, politically, and militarily. Without satellites, many of our factories and distribution systems would break down and our "just in time" inventory systems would at least temporarily stall, interrupting the flow of food, medicines, and commodities to our stores and markets. Without spy satellites, our military would be operating blind, thereby increasing the likelihood of conflict through fear and opportunism by our potential adversaries. We have become dependent upon space – and we are mostly ignorant of that fact.

The exploration and use of space began in 1957 with the launch of Sputnik 1 by the Soviet Union. Since then, the nations of the world have launched over 8,000 spacecraft. Of these, several hundred are still providing information and services to the global economy and the world's governments. Over time, nations, corporations, and individuals have grown accustomed to the services they provide and many are dependent upon them. Commercial aviation, shipping, emergency services, vehicle fleet tracking, financial transactions, and agriculture are areas of the economy that are increasingly reliant on space.

For example, the Global Positioning System (GPS), a network of satellites in low-Earth orbit (LEO) developed to provide precise position information to the military, is now in common use by individuals and industry. Another example is communications. Since Telstar-1 initiated worldwide television transmission in 1962, now-ubiquitous cable television signals are sent to local providers via satellite relay. The science of weather forecasting is now totally dependent upon continuous real-time information from multiple satellites circling the globe. These are but a few examples of areas in which we now depend upon space technology.

As we've grown dependent upon space, so have we been sowing the seeds of a problem that might ultimately deny it to us. Dead satellites, spent rocket stages, and bits of debris from a satellite deliberately obliterated in orbit make up a sizable portion of an estimated 500,000 pieces of debris now circling the Earth – any one of which is traveling fast enough to destroy a functioning satellite if it were to collide with one. According to the experts, we are dangerously close to the point at which a runaway series of collisions could pump so much debris into orbit as to make it essentially unusable.

Unfortunately, space is now a potential area for conflict. The detonation of one or more nuclear weapons in Earth orbit could have two disastrous consequences for inadequately protected satellites. First is the immediate damage to spacecraft electronics from the high-energy electromagnetic pulse generated by the bomb. Without functioning electronics, a spacecraft becomes an orbiting piece of junk. The second is the longer-term degradation and loss of spacecraft electronics resulting from repeatedly passing through the artificial radiation belts that would be created from the detonation of a nuclear bomb in orbit.

The threats to our satellites are not all made by the human race. The Earth periodically experiences geomagnetic storms. These storms occur when a solar wind shock wave strikes the Earth's magnetic field after some increase in solar activity like a flare. Associated with these storms are dramatically increased radiation levels which are potentially damaging to spacecraft and their electronics.

Part 1: How We Might Lose Our Satellites

Chapter 1: Orbital Debris
The human-caused orbital debris problem is bad and getting worse. There are currently over 500,000 nonfunctional spacecraft and pieces of junk orbiting the Earth at five miles per second or faster. Despite the fact that the space around the Earth is big, pieces of debris do collide with each other and with functioning spacecraft. When a collision happens, many more pieces of junk are created, increasing the risk for the remaining satellites. In early 2009, two satellites collided with a relative speed of over 26,000 miles per hour. Both satellites were destroyed and the debris cloud, consisting of hundreds of individual pieces, quickly spread out around the Earth. Each new piece of debris is now a threat to other spacecraft. Studies indicate that a cascade effect of collisions forming debris, causing yet more collisions and more debris, etc., could result in LEO becoming unusable within the next 50 years.

Chapter 2: Space War
In 2007, China destroyed a satellite in space and significantly increased the Earth-orbiting debris population. Shortly thereafter, the United States did the same, albeit in a manner that did not significantly increase the orbital debris problem. It is possible that an emerging space power, any country with the capability of launching a satellite into space, could intentionally collide strategically targeted rockets with a few operating spacecraft, causing a cascade of debris formation.

Rogue states with nuclear weapons pose another risk to our space infrastructure. Experiments in the 1960s showed that the detonation of a nuclear weapon in LEO might produce an artificial radiation belt around the planet that would persist for years, potentially damaging or destroying a significant number of satellites as they pass through it. Alternatively, detonating a few nuclear weapons optimized to produce intense bursts of radio noise called an electromagnetic pulse (EMP) could knock out satellites' electronics almost immediately.

Chapter 3: Solar Storms

Solar storms happen and, when they do, the effects on satellites can be devastating. For example, two Canadian telecommunications satellites were disabled by a solar storm in 1994. The first recovered in just a few hours. The other didn't recover for six months.

The Space Age began about 50 years ago. We've been monitoring solar activity for longer than that, but we have no idea whether the space radiation conditions we've enjoyed since the dawn of the Space Age are the norm or whether the relatively benign conditions we've experienced will end tomorrow with a more active Sun sending intense radiation into space – frying our satellites in the process.

Part 2: If We Were to Lose Our Satellites ...

Chapter 4: The Global Positioning System (Military Uses)

Since 1995, a network of between 24 and 32 American satellites have been orbiting the Earth, providing continuous reliable position and navigation services to users around the world. Built to support the needs of the US military, the system is now widely used by individuals and commercial companies for navigation, surveying, tracking shipments and commerce, and many other applications that require precise timing and location information. It has been so successful that other countries are now building their own satellite systems to provide similar services.

Many consider the Gulf War of 1990–91 to be the world's first war won using space. America and its allies used space reconnaissance, GPS, and GPS-guided munitions to rapidly decimate the Iraqi Army. America's former Cold War adversaries had supplied Iraq with arms and the brief conflict showed that the application of space technology was a critical element in their rapid defeat. Soldiers, tanks, aircraft, and ships use GPS to know where they are. GPS-guided bombs struck with precision unrivaled in the history of warfare. If an adversary were to strike when we're scrambling to operate without GPS, how well would we be able to respond?

Chapter 5: Economic Fallout

According to the US Department of Transportation, there are more than 10 million trucks in the United States alone. A large fraction of these trucks use GPS to help them navigate from the factory to the store, from the supplier to the manufacturer, or from the producer to the consumer. The companies which own many of these trucks, or the companies that contract with the independent owners, gather data from the GPS devices on board the trucks to help them manage the delivery of food to the grocery stores, raw materials to the factories, and medical supplies to the pharmacies and hospitals. If the satellites were to go off line, this complex chain of food, raw materials, and supplies would break – with serious consequences.

Chapter 6: The Global Positioning System and the Average Person
If you need help navigating in a strange city, you are likely to use your GPS receiver. For under $200, you may purchase a fully functional GPS receiver that will help you get safely from Point A to Point B. They are now a standard option in many rental cars. Have you bought a house or some land recently? Chances are that a survey was performed using GPS's ability to provide highly accurate position knowledge as a part of the process.

Scientists use GPS data in their study of earthquakes and of the Earth's atmosphere. It is difficult to find a realm of modern life that is untouched by GPS. How would we adapt if GPS were to disappear?

Chapter 7: Spy Satellites and Military Communications
Modern satellite reconnaissance is used for all aspects of military information gathering – detection and imaging of adversary command, mobilization, and training areas; tracking of ground forces, ships, and aircraft; and intercepting communications. Much of the leverage available to American and European troops on the modern battlefield is due to the advantages of military reconnaissance. If this advantage were lost, troops in the field would lose both strategic and tactical superiority, resulting in longer and more protracted battles, greater casualty rates, and less chance of achieving battlefield advantage. And, without satellite tracking and communications, the logistics of supporting combat troops around the world would be nearly impossible.

Chapter 8: Communications
Immediately after the events of 9/11, the number of cell phone calls spiked. Today, cell phone companies use GPS timing signals to help them quickly and efficiently route calls. Without GPS, cell phones might become paperweights. Retail stores, gasoline stations, and other businesses use satellites to approve credit card transactions.

Where do most people go for their up-to-date information in a crisis? Television. If our communications satellites go down, national and international news channels will go down with them – as do all satellite television and cable television stations.

In addition to satellite-guided munitions and real-time satellite imagery, the United States' military advantage depends upon reliable global communications. It is estimated that more than 75% of all military communications are transmitted via satellite. And if those satellites were to go silent ...

Chapter 9: Weather Forecasting
Once upon a time, in Galveston, Texas, there came a hurricane. To the citizens of Galveston, the day began much like any other except for the gathering storm clouds on the horizon. By the end of the day, thousands of people were trapped on an island that was about to be ravaged by a hurricane that no one saw coming. Eight thousand of them died. This was in 1900.

Once upon a more recent time, in that same city, residents knew of a coming storm called "Ike" long before it made landfall in their beautiful city. Satellites and aircraft tracked the storm since its formation off the coast of Africa and with this information, forecasters were able to warn the residents of Galveston of the impending storm with enough lead time for all who were in harm's way to evacuate. Had the satellites failed, how many aircraft would it have taken to continuously monitor the 29-million-square-mile Atlantic Ocean during the six-month hurricane season in order to prevent a repeat of what happened in 1900?

And hurricanes are far from being the only reason we rely on our satellites for weather forecasting ...

Chapter 10: Remote Sensing: Environmental Monitoring and Science
For over 30 years, we've been able to monitor global climate patterns and changes using satellite remote sensing. Much of what we've learned about the cycles that drive our climate and our technological civilization's impact on the global ecosystem has come from satellite observations. Without updated information from space, we would be crippled in our ability to monitor atmospheric changes, global rainfall patterns, and other climatological indica-tors – leaving policy makers to make decisions without the most significant part of their data in hand.

Satellite systems make regional and global resource monitoring possible. This is because it is very difficult and costly to conduct ground and aerial surveys over large areas and then to coordinate the individual surveys by joining them together. To collect data on a global scale, one must use the unique vantage point provided by space systems. One of the most successful applications of space imaging is monitoring the world's agricultural production, including identifying and differentiating most of the major crop types: wheat, barley, millet, oats, corn, soybeans, rice, and others. Feeding the world is only possible because of our ability to monitor food production and rapidly adapt to changes in the distribution system – and, in our modern world, both of these require space satellite systems.

Satellite remote sensing has also been successfully used in identifying mineral resources, particularly when the data from various types of space-based sensors are combined and compared. Locating future sources of raw materials suddenly becomes much more difficult and costly without satellite data.

Chapter 11: The International Space Station and Human Space Flight
Any event that creates a serious hazard for unmanned satellites in space creates a direct threat to the life of those on board manned spacecraft or space stations. The International Space Station (ISS) has been continuously occupied for over nine years and its crew size was recently increased to six. What would happen to the station and its crew if some event caused it to become uninhabitable?

Chapter 12: Effects on Scientific Research Satellites
The loss of such assets as the Hubble Space Telescope or the Chandra X-Ray Observatory may not directly impact the lives of people on the Earth, but we

would be inestimably poorer without the breadth and depth of knowledge about the universe that we obtain from them.

Part 3: What Can We Do?

Chapter 13: Reduce the Growth in Orbital Debris
Fortunately, the world's spacefaring nations are aware of the growing threat posed by orbital debris and are taking steps to prevent the problem from getting worse. Unfortunately, the problem is already serious and it will not get better on its own. Most countries now require that new spacecraft either de-orbit or move out of the way when their missions are completed. Scenarios and approaches for reducing the growth of orbital debris are presented.

Chapter 14: Reduce the Amount of Debris in Space
Technologies to remove some of the debris circling the Earth are being assessed and systems to actually begin removing space junk may come on line within the next decade or two:

- space and ground-based lasers might push smaller pieces of junk into orbits that will more quickly decay, causing them to burn up in the Earth's atmosphere;
- for mid-sized orbital debris, such as nonfunctional satellites, a small propulsive device that attaches to the "dead" spacecraft and moves it out of the way is required;
- large pieces of debris, like spent rocket stages, will require much larger propulsive devices to be attached; in some cases, it may make more sense to boost the debris to a higher orbit where it is not at high risk of impacting another satellite (instead of causing it to enter the Earth's atmosphere and burn up).

Chapter 15: Harden against Space Radiation (Contributed by Dr James K. Woosley)
Some amount of hardening against space radiation is already incorporated into satellite design. However, as we learn more about the space radiation environment and particularly solar flares, the risk appears to be greater than was first supposed. Future generations of satellites must be shielded or include protective measures against the worst-case radiation loads of the most intense solar flare foreseeable. This hardening will also be helpful in the worst-case event of a nuclear detonation in space.

Appendix A: More on Orbital Debris
From close calls and near misses to the rapid increase in the number of debris objects, the orbital debris problem is technically complex and legally difficult to address.

Appendix B: The Kessler Effect as Originally Described
The seminal paper from Dr Don Kessler, as originally published in the *Journal of Geophysical Research*, is a call to arms for dealing with the growing orbital debris problem.

Appendix C: (Selected) Spacecraft Failures and Anomalies Due to Solar and Geomagnetic (Solar Event-Induced) Storms or Orbital Debris Impacts
A list of selected spacecraft involved in solar event-induced failures and anomalies with a brief description of each event.

Appendix D: International Agreements Governing Orbital Debris and Space Weaponization
A look at the legal implications for orbital debris removal and space weaponization with reference to international agreements and treaties.

Introduction

It might begin with the accidental collision between two orbiting satellites – a rare occurrence, but it has happened, most recently in 2009. It might be the opening shot of new kind of war, begun with a foreign power destroying a critical spy satellite – China demonstrated the capability when they destroyed an aging weather satellite in 2007; the United States followed suit only a few months later. It might even be caused by a large blast of radiation coming from the Sun causing a spacecraft to malfunction and crash into another.

Whatever the initial cause, the result may be the same. A satellite destroyed in orbit will break apart into thousands of pieces, each traveling at over 8 km/sec. This virtual shotgun blast, with pellets traveling 20 times faster than a bullet, will quickly spread out, with each pellet now following its own orbit around the Earth. With over 300,000 other pieces of junk already there, the tipping point is crossed and a runaway series of collisions begins. A few orbits later, two of the new debris pieces strike other satellites, causing them to explode into thousands more pieces of debris. The rate of collisions increases, now with more spacecraft being destroyed. Called the "Kessler Effect", after the NASA scientist who first warned of its dangers, these debris objects, now numbering in the millions, cascade around the Earth, destroying every satellite in low-Earth orbit.

Without an atmosphere to slow them down, thus allowing debris pieces to burn up, most debris (perhaps numbering in the millions) will remain in space for hundreds or thousands of years. Any new satellite will be threatened by destruction as soon as it enters space, effectively rendering many Earth orbits unusable. But what about us on the ground? How will this affect us?

Imagine a world that suddenly loses all of its space technology. If you are like most people, then you would probably have a few fleeting thoughts about the Apollo-era missions to the Moon, perhaps a vision of the Space Shuttle launching astronauts into space for a visit to the International Space Station (ISS), or you might fondly recall the "wow" images taken by the orbiting Hubble Space Telescope. In short, you would know that things important to science would be lost, but you would likely not assume that their loss would have any impact on your daily life.

Now imagine a world that suddenly loses network and cable television, accurate weather forecasts, Global Positioning System (GPS) navigation, some cellular phone networks, on-time delivery of food and medical supplies via truck and train to stores and hospitals in virtually every community in America, as well as science useful in monitoring such things as climate change and agricultural sustainability. Add to this the crippling of the US military who now depend upon spy satellites, space-based communications systems, and GPS to know where their troops and supplies are located at all times and anywhere in the world. The result is a nightmarish world, one step away from nuclear war, economic disaster, and potential mass starvation. This is the world in which we are now perilously close to living.

Space satellites now touch our lives in many ways. And, unfortunately, these satellites are extremely vulnerable to risks arising from a half-century of carelessness regarding protecting the space environment around the Earth as well as from potential adversaries such as China, North Korea, and Iran.

No government policy has put us at risk. It has not been the result of a conspiracy. No, we are dependent upon them simply because they offer capabilities that are simply unavailable any other way. Individuals, corporations, and governments found ways to use the unique environment of space to provide services, make money, and better defend the country. In fact, only a few space visionaries and futurists could have foreseen where the advent of rocketry and space technology would take us a mere 50 years since those first satellites orbited the Earth. It was the slow progression of capability followed by dependence that puts us at risk.

The exploration and use of space began in 1957 with the launch of Sputnik 1 by the Soviet Union. The United States soon followed with Explorer 1. Since then, the nations of the world have launched over 8,000 spacecraft. Of these, several hundred are still providing information and services to the global economy and the world's governments. Over time, nations, corporations, and individuals have grown accustomed to the services these spacecraft provide and many are dependent upon them. Commercial aviation, shipping, emergency services, vehicle fleet tracking, financial transactions, and agriculture are areas of the economy that are increasingly reliant on space.

Telestar 1, launched into space in the year of my birth, 1962, relayed the world's first live transatlantic news feed and showed that space satellites can be used to relay television signals, telephone calls, and data. The modern telecommunications age was born. We've come a long way since Telstar; most television networks now distribute most, if not all, of their programming via satellite. Cable television signals are received by local providers from satellite relays before being sent to our homes and businesses using cables. With 65% of US households relying on cable television and a growing percentage using satellite dishes to receive signals from direct-to-home satellite television providers, a large number of people would be cut off from vital information in an emergency should these satellites be destroyed. And communications satellites relay more than television signals. They serve as hosts to corporate video conferences and convey business, banking, and other commercial information to and from all areas of the planet.

The first successful weather satellite was TIROS. Launched in 1960, TIROS operated for only 78 days but it served as the precursor for today's much more long-lived weather satellites, which provide continuous monitoring of weather conditions around the world. Without them, providing accurate weather

Figure I.1 Hurricane Ivan, as seen by NASA's Terra satellite in 2004. (Image courtesy of NASA)

forecasts for virtually any place on the globe more than a day in advance would be nearly impossible. Figure I.1 shows a satellite image of Hurricane Ivan approaching the Alabama Gulf coast in 2004. Without this type of information, evacuation warnings would have to be given more generally, resulting in needless evacuations and lost economic activity (from areas that avoid landfall) and potentially increasing loss of life in areas that may be unexpectedly hit.

The formerly top-secret Corona spy satellites began operation in 1959 and provided critical information about the Soviet Union's military and industrial capabilities to a nervous West in a time of unprecedented paranoia and nuclear risk. With these satellites, US military planners were able to understand and assess the real military threat posed by the Soviet Union. They used information provided by spy satellites to help avert potential military confrontations on numerous occasions. Conversely, the Soviet Union's spy satellites were able to observe the United States and its allies, with similar results. It is nearly impossible to move an army and hide it from multiple eyes in the sky.

Satellite information is critical to all aspects of US intelligence and military planning. Spy satellites are used to monitor compliance with international arms treaties and to assess the military activities of countries such as China, Russia, Iran, and North Korea. Figure I.2 shows the capability of modern *unclassified* space-based imaging. The capability of the classified systems is presumed to be significantly better, providing much more detail. Losing these satellites would place global militaries on high alert and have them operating, literally, in the blind.

Our military would suddenly become vulnerable in other areas as well. GPS, a network of 24–32 satellites in medium-Earth orbit, was developed to provide precise position information to the military, and it is now in common use by individuals and industry. The network, which became fully operational in 1993, allows our armed forces to know their exact locations anywhere in the world. It is used to guide bombs to their targets with unprecedented accuracy, requiring that only one bomb be used to destroy a target that would have previously required perhaps hundreds of bombs to destroy in the pre-GPS world (which, incidentally, has resulted in us reducing our stockpile of non-GPS-guided munitions dramatically). It allows soldiers to navigate in the dark or in adverse weather or sandstorms. Without GPS, our military advantage over potential adversaries would be dramatically reduced or eliminated.

GPS is also used in our civilian economy. Aside from it being a handy way to navigate in a strange city using the portable GPS receivers so many of us have in our personal and rental cars, it is also used to guide emergency rescue personnel to the scenes of accidents, fires, and other emergencies. For example, many local fire departments now rely on GPS data to guide their fire trucks and other emergency response vehicles to their destinations. They are used to show firefighters where to go and, once they arrive at their destination, the location of nearby water hydrants. In my adopted hometown of Madison, Alabama, the city no longer maintains reflectors in the roadway showing the location of water hydrants – they now totally rely on GPS data.

Figure I.2 Boston, Massachusetts, as seen from space. (Image courtesy of GeoEye)

In the future, the current beacon-based commercial aircraft navigation system is to be replaced by GPS. With tens of thousands of commercial aircraft in the sky at any one time, a sudden loss of a critical navigation component (like knowing where you are) would result in chaos – especially if there were no backup plan in place to implement in this type of emergency. Commercial aviation, for both people and cargo, would grind to a halt.

Credit card companies increasingly use satellite relays to transmit approval and purchase information from stores to their central data-processing facilities.

Many grocery stores, gas stations, and other retailers now have satellite dishes on their rooftops to allow fast communication, via satellite, with the credit card companies. Trucking companies use GPS to know the location of shipments traveling from warehouses to retail stores, allowing better inventory management and preventing shortages resulting from less efficient inventory management. Without GPS, today's system of overnight package delivery would grind to a halt, crippling many businesses that rely on rapid distribution of their products.

Governments use satellite data to make decisions regarding development, resource allocation, as well as the politically sensitive topics of global warming and climate change. It is from global satellite information that we are able to nearly continuously monitor the amount of ice in the Arctic and Antarctic, the change in global sea levels and sea-surface temperatures, and changes in atmospheric temperatures. While the loss of these data would not be of immediate consequence to the person on the street, international policy makers would be unable to make informed decisions on matters of trade and pollution control.

The United States now depends upon space technology economically, politically, and militarily. Without satellites, our factories and distribution systems would break down and our "just in time" inventory systems would stall, interrupting the flow of food, medicines, and commodities to our stores and markets. Without spy satellites, our military would be operating blind, thereby increasing the likelihood of conflict facilitated by fear and opportunism among our potential adversaries. We are now dependent upon space – and we are mostly ignorant of that fact.

As we've grown dependent upon space, so have we been sowing the seeds of a problem that might ultimately deny it to us. Dead satellites, spent rocket stages, and bits of debris from a satellite deliberately obliterated in orbit make up a sizable portion of an estimated 500,000 pieces of debris now circling the Earth (Figure I.3) – any one of which is traveling fast enough to destroy a functioning satellite in a collision.

In 2007, a piece of orbital debris hit one of the Space Shuttle's radiator panels and made a hole three-tenths of an inch wide – if it had collided with a critical system, the results could have been catastrophic (Figure I.4). According to the experts, we are dangerously close to the point at which a runaway series of collisions could pump so much debris into orbit as to make it unusable. The Kessler Effect, mentioned earlier, is of grave concern to space agencies worldwide and many departments of defense.

In addition to the consequences for us here on the Earth, the immediate impact (pun intended) to astronauts on the ISS would potentially be their death. The ISS already performs several maneuvers each year to avoid potential collisions with known pieces of debris. If the debris population were to increase rapidly, as would be the case with the Kessler Effect, the ISS might suffer a series of crippling collisions, resulting in its destruction. It remains to be seen whether or not the astronauts on board would have sufficient warning time to escape.

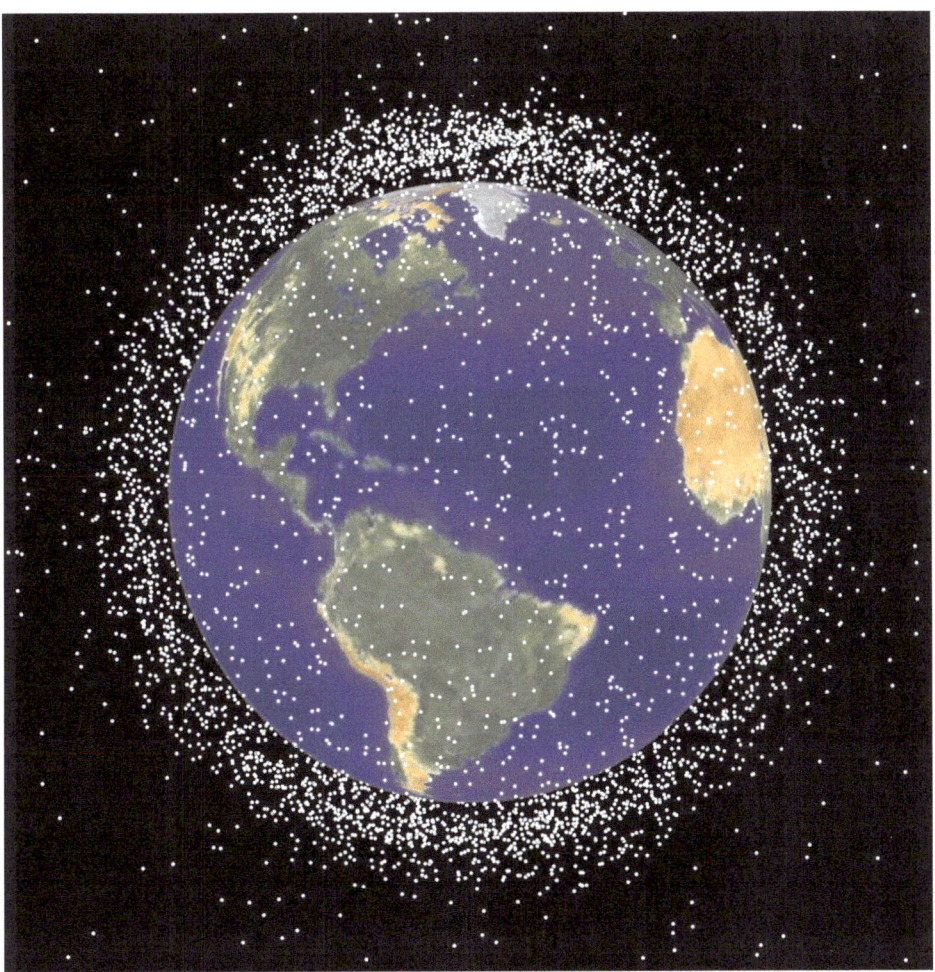

Figure I.3 The Earth is surrounded by a cloud of spacecraft and debris that number hundreds of thousands. (Image courtesy of NASA)

Why have we let this happen? That's a good question. Scientists have known since the beginning of the Space Age that an object placed in a high enough orbit will remain there, orbiting the Earth, for centuries or millennia. Without any outside forces acting to slow it down, such as an atmosphere, old satellites will continue to circle the globe at tremendous speeds. Many of these satellites were not designed to remain intact much past their estimated operational lifetime. Older satellites contain pressurized tanks that eventually fail, potentially causing explosions that turn one large piece of debris traveling at 8 km/sec into hundreds of pieces, each traveling at 8 km/sec. To make matters worse, the rockets used to take spacecraft to space often were not designed to re-enter the Earth's atmosphere and burn up. Rather, they were simply cast aside

Figure I.4 Entry-hole damage to the Space Shuttle's radiator panel caused by a collision with a small piece of orbital debris. (Image courtesy of NASA)

and forgotten. Many of the dead rockets suffered the same fate as the satellites they launched; they blew up or broke apart into hundreds of debris objects. Some satellites were even deliberately destroyed to prevent them from being recovered by potential adversaries.

Unfortunately, space is now a possible area for conflict. The detonation of one or more nuclear weapons in Earth orbit could have two disastrous consequences for inadequately protected satellites. First is the immediate damage to spacecraft electronics from the high-energy electromagnetic pulse generated by the bomb. Without functioning electronics, a spacecraft becomes an orbiting piece of junk. The second is the longer-term degradation and loss of spacecraft electronics resulting from repeatedly passing through the artificial radiation belts that would be created from the detonation of a nuclear bomb in orbit.

The military threat to our satellites is not all due to nuclear weapons. In January 2007, China launched a missile into space and destroyed one of their old weather satellites. This was their way of announcing to the world that they are now a space power capable of destroying anyone else's satellites in space. The satellite broke apart, creating over 2,000 pieces of debris [1]. This single act was the largest debris-generating event in space history. With this test, China increased by 40% the probability that the Hubble Space Telescope will be hit by a piece of debris!

Any country that can launch a satellite into space is not far from having the capability of destroying satellites as well. If you can imagine the impact of blowing up a satellite containing thousands of ball bearings, then you can understand how a relative newcomer in space might disable or destroy all of the satellites in any given orbital region. For reference, Iran launched its first orbiting space satellite in 2009.

The threats to our satellites are not all made by humans. The Earth periodically experiences geomagnetic storms. These storms occur when a solar wind shock wave strikes the Earth's magnetic field after some increase in solar activity like a flare or coronal mass ejection. Associated with these storms are dramatically increased radiation levels that are potentially damaging to spacecraft and their electronics. In 1994, two Canadian satellites were damaged by one of these solar storms [2]. As a result, some satellite-to-ground communications, including television broadcasts, were knocked out for five months.

There is virtually nothing we can do to protect our spacecraft from the effects of being hit by a piece of debris traveling 23 times faster than a bullet fired from a 9-mm pistol. A piece of debris that hits a spacecraft will either damage it or destroy it. However, there are actions we can take now to reduce the likelihood of losing our satellites in the near future.

First of all, the space powers of the world must build into their future spacecraft the capability of being removed from orbit after their mission is complete. This will reduce the rate at which the problem is growing. The United States requires that all of its government-funded satellites be able to de-orbit within 25 years of their mission ending. Some meet this requirement by being in a low enough orbit that residual atmospheric drag will gradually cause the spacecraft's orbital altitude to drop, eventually resulting in its burning up in the atmosphere. Spacecraft placed in higher orbits can meet the requirement by adding an additional rocket motor that will be activated at the end of the mission, forcing the spacecraft into a lower orbit and eventual burn-up.

Next, a series of missions to capture and remove old, dead satellites from orbit must be built and flown. Finally, a method of capturing some of the 500,000 or more small pieces of debris must be developed and implemented. This is not an easy task. The debris is tiny; many pieces are no larger than a sugar cube. In addition, the debris is scattered in various orbits which are mostly populated by empty space. Developing a spacecraft that can eliminate the debris in the right place and at the right time will be a challenge.

To protect against the radiation effects of a massive solar storm or multiple nuclear detonations in space will require adding mass to the spacecraft and designing spacecraft systems to have more redundancy. Most military satellites already do this but very few commercial spacecraft customers are willing to pay the extra costs required.

Our global economy is now so dependent upon satellites that their loss would be devastating. This dependence is not yet global. For now, the countries with the most developed economies are the most at risk: the United States, Japan, and Europe. Unfortunately, we in the developed world are therefore vulnerable as never before to either the runaway Kessler Effect or the hostile actions of a smaller and less capable adversary – setting us up for a potential "Pearl Harbor in space". Some policy makers and defense planners are aware of the threat and, for the first time, there appears to be some sort of international awareness and a sense of urgency for doing something about it. For the most part, we are blissfully ignorant of how much space has become intertwined with our lives and we are likely to remain so – *unless something happens that causes us to live on an Earth without satellites*.

References

[1] Kelso, T.S. Analysis of the 2007 Chinese ASAT Test and the Impact of Its Debris on the Space Environment. Proceedings of the 2007 AMOS Conference, Maui, Hawaii.
[2] Beddingfield, K.L.; Leach, R.D.; Alexander, M.B. *Spacecraft System Failures and Anomalies Attributed to the Natural Space Environment*. NASA Reference Publication 1390, August 1996.

Part 1

How We Might Lose Our Satellites

1 Orbital Debris

Imagine being hit with a bowling ball moving at 100 km/hr. Now imagine being hit with pea traveling at 7 km/sec (25,200 km/hr). Both impacts would release about the same amount of energy, causing comparable damage. By anyone's definition, this would make it a bad day. When it comes to being hit by a moving object, size matters – but speed matters more.

There are more than 500,000 pieces of pea-sized and larger junk orbiting the Earth at these tremendous speeds. Most of this junk is in a region of space below about 2,000 km in altitude, commonly known as low-Earth orbit (LEO). Space junk is not confined to any single orbit, although there are some altitude regions that have more junk than others [1]. When one of these pieces of junk collides with a satellite, the results can be catastrophic.

Space debris comes in many shapes and sizes. From spent rocket stages several meters across, to 1-cm globs of radioactive coolant from old nuclear-powered satellites, to bits of paint no more than a fraction of a centimeter in diameter, the population of space debris objects is growing larger every year. When a spacecraft is hit by a piece of debris, the effects can range from minor to devastating. As you might suspect, this depends upon the size and speed of the debris that hits it.

Figure 1.1 shows just how dangerous even a small piece of debris can be when it is traveling at 7 km/sec. A large block of aluminum, the same material from which many satellites are made, was the target for a small aluminum ball in a laboratory test conducted by the European Space Agency (ESA). The ball used in the test was completely vaporized when it collided with the block, producing the crater shown in the figure.

How did this relatively small aluminum ball cause so much damage? The answer is: speed. The damage caused during an impact is mostly determined by the energy of the collision. Since an object's kinetic energy, or energy of motion, is determined by both its mass and its velocity, the faster an object is moving, the more energy it will release – causing more damage. The problem is that the energy increases as the square of the velocity; doubling the mass also increases an object's kinetic energy, but only by a factor of two.

How did we let space junk become the problem that it is today?

When the Soviet Union launched Sputnik in 1957, they also placed the first debris into orbit when they discarded the launch vehicle's upper stage. More junk likely entered space when the United States launched Explorer 1 three months later. Since then, the countries of the world have launched thousands of rockets into space – and the population of orbiting debris has gone up with each launch. One of the earliest spacecraft of the Space Age, Vanguard 1, launched in

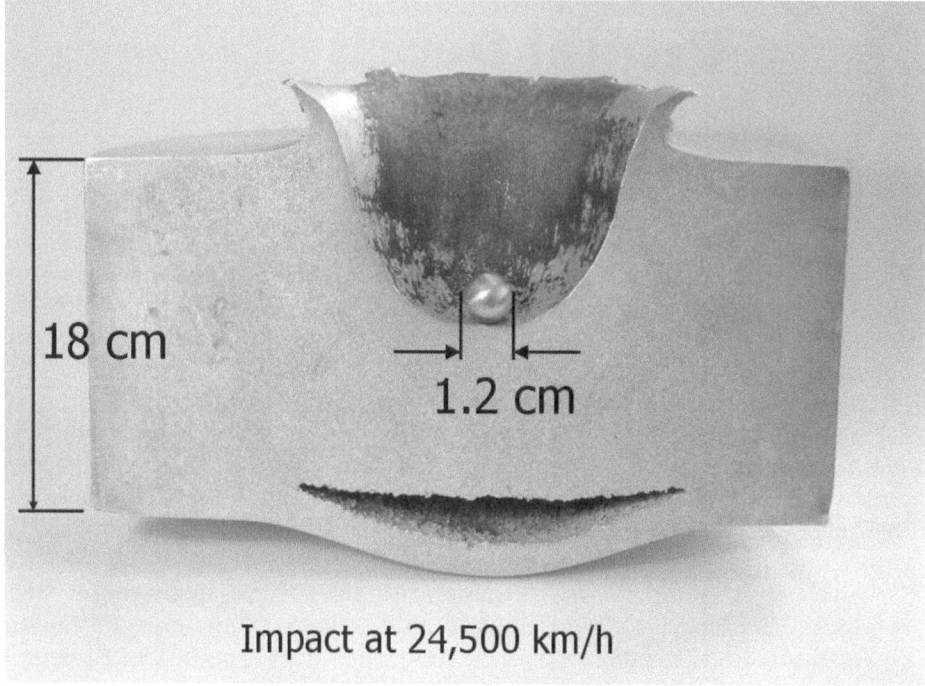

Figure 1.1 This image shows the result of an experiment in which a small aluminum ball traveling at orbital velocity hit a block of aluminum, causing the damage crater seen in the center of the aluminum block. The original ball was vaporized in the collision. The ball shown in the figure was placed inside the crater for size comparison. (Image courtesy of ESA)

1958, is still in orbit today and it will likely remain there for another 200 years. Vanguard 1 stopped working in 1964 [2].

Unfortunately, the extent of the debris problem we have today was not directly caused by the launch of these satellites. If each had contributed only a few pieces of insulation, paint, or the accidental bolt to the debris population, then we would have only a few thousand pieces of junk to worry about.

The majority of debris in Earth orbit today was created by the explosion of spacecraft or the rockets used to put them into space [3]. For example, over 200 satellites or launch vehicle upper stages have detonated in space, each event contributing hundreds or thousands of debris objects to the total population. We are all too familiar with the image of the Space Shuttle *Challenger* exploding as it launched towards space on that fateful day in 1986 (Figure 1.2). Similar events, fortunately on unmanned rockets, sometimes happen in space – far from the cameras. And it is these explosions that dramatically increase the number of debris objects in space.

Some of these explosions were caused by rocket failures during or shortly after launch. Many occurred months or even years later as the pressurized gas tanks in

Figure 1.2 When the Space Shuttle *Challenger* exploded, the debris cloud fell quickly to the ground. In space, each piece would have gone into orbit around the Earth. (Image courtesy of NASA)

a discarded rocket body or satellite failed, with the same end result. For example, in February 2006, a Proton rocket malfunctioned, placing itself into Earth orbit as one large piece of debris. One year later, the rocket exploded and created a debris cloud consisting of more than 1,000 objects [4].

Failed batteries are a common cause of satellite explosions. Virtually any battery can fail under certain circumstances and, in the hostile environment of space, many have done so. One has to look no further than the laptop computers many of us use to be aware of this problem. Over the last few years, computer manufacturers have had to recall batteries that were sometimes exploding and injuring people. Even with rigorous testing and the highest possible engineering standards, batteries can still fail – causing damage when they do so.

Some explosions may have even been intentional. While we may never know for sure, the former Soviet Union is thought to have intentionally destroyed as many as 50 spacecraft. I am speculating that this may have been in part a result of the US Space Shuttle becoming operational. The Shuttle was designed to be able to retrieve satellites in space and there is no fundamental reason why it

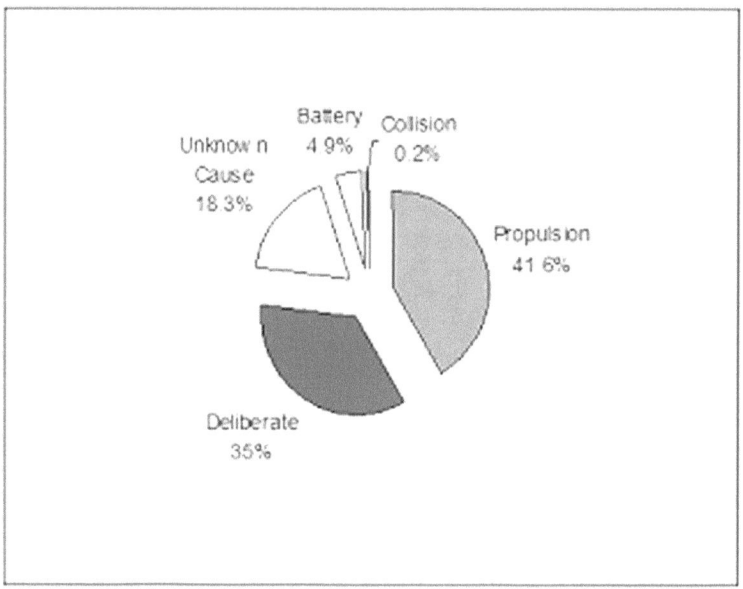

Figure 1.3 Percentage of debris break-up object causes according to NASA in 2007. (Image courtesy of NASA)

could not have "retrieved" someone else's satellite. If I were that someone, then I would certainly want to make it known that such a retrieval could have devastating consequences. NASA estimates that 35% of known satellite break-ups were intentional (Figure 1.3).

Fragmentation also occurs in space when two objects collide. While space is, quite literally, big, it is not so big as to make it impossible for two spacecraft to run into each other. The 2009 collision between the fully functioning Iridium 33 and the long-non-operational Cosmos 2251 resulted in the creation of more than 2,000 rather large pieces of debris and an estimated 100,000 small pieces [5]. It is important to note that the debris created by this and other collisions doesn't remain localized. Each piece of debris was placed into a new and unique orbit by the energy imparted to it during the collision. Not only were pieces flung into both higher and lower altitudes, but, as indicated by their commensurately different new orbital periods, the pieces were flung into totally new orbits (Figure 1.4). In other words, pieces of the two satellites didn't remain in the orbits that the satellites had been in before the collision; some of the debris entered orbits that directly threaten the Hubble Space Telescope and the International Space Station.

There have also been satellites hit by small pieces of debris, causing everything from minor damage to a complete loss of the satellite – creating yet more debris. In 1996, the French military satellite, Cerise, is thought to have been struck by a piece of debris originating from the fragmentation of a European Ariane rocket.

Not all collisions are accidental. In 1985, the United States tested an anti-

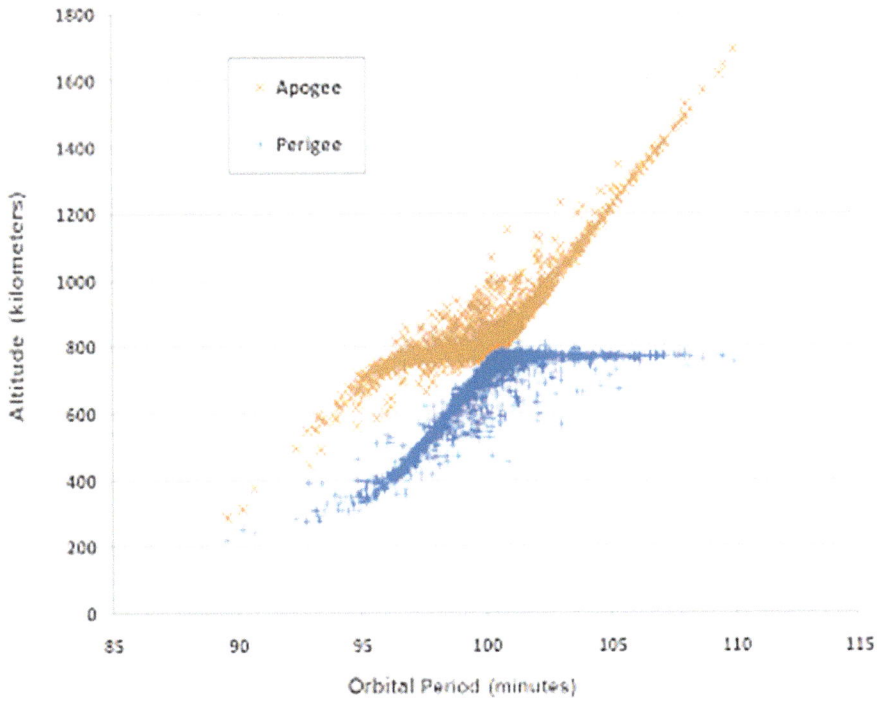

Figure 1.4 Altitude distribution of the debris created by the collision of Iridium and Cosmos satellites in 2009. (Image courtesy of NASA)

satellite weapon by destroying the Solwind satellite in LEO. The resulting debris burned up in the atmosphere within the next 20 years.

In 2007, the Chinese government tested an anti-satellite weapon by deliberately destroying an aging weather satellite, Fengyun-1C, creating at least 3,000 large pieces of debris and an estimated 100,000 small pieces (Figure 1.5) [6]. Unfortunately, the orbit of Fengyun-1C was such that the debris cloud created will likely remain for many decades or longer. The significance of this deliberate act cannot be understated. The debris cloud resulting from this anti-satellite test now accounts for more than 25% of all the cataloged objects in LEO and increased the risk of the Hubble Space Telescope being hit by 40%!

The United States used one of its ASAT weapons in 2008 to destroy satellite USA-193. Fortunately, all but one cataloged piece of debris re-entered and burned up within months of being created.

Figure 1.6 shows the growth of orbital debris over time. As you can see, the number of debris objects was increasing slowly until the 2007 Chinese ASAT test and the 2009 Iridium/Cosmos collision. Those two events nearly doubled the orbital space junk population [7].

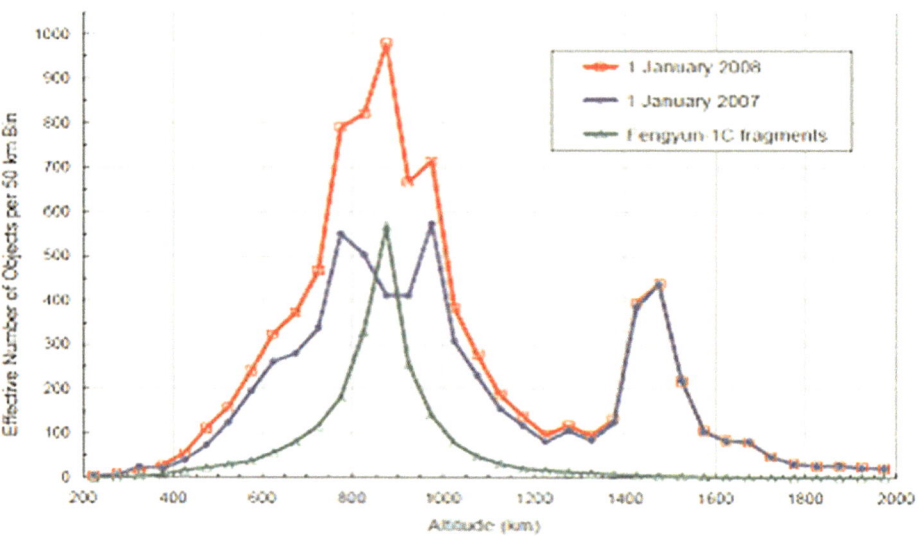

Figure 1.5 The debris caused by the destruction of the Fengyun-1C satellite (the smallest sharp peak centered on about 900 km) increased the total number of debris objects in LEO by 1 January 2008 (the top curve) dramatically from the 2007 levels – the middle curve. (Image courtesy of NASA)

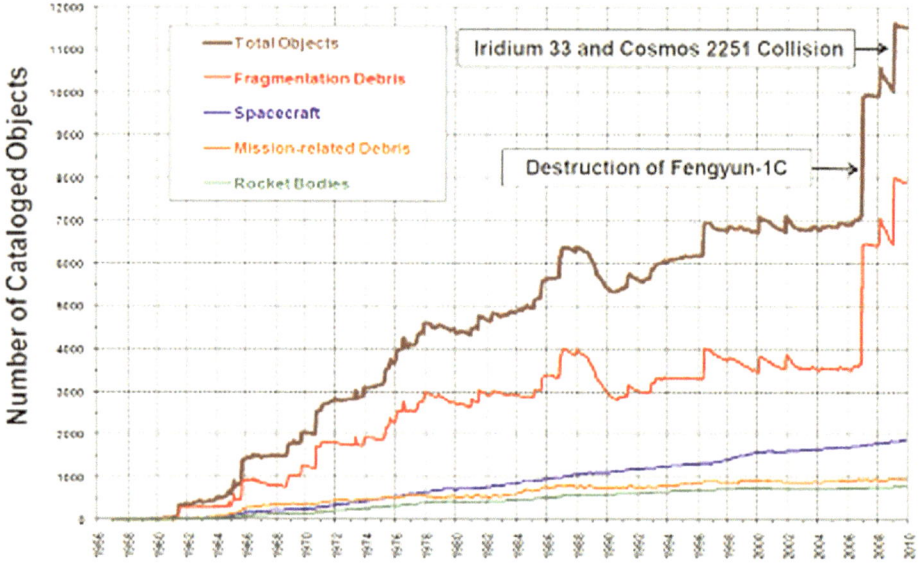

Figure 1.6 The graph shows the dramatic increase in the number of orbital debris objects following the Chinese ASAT test of 2007 and the Iridium/Cosmos collision in 2009. (Image courtesy of NASA)

Table 1.1. Top 10 debris-causing break-ups as of May 2010 [8].

Name	Event year	Cataloged debris objects	Debris-causing event
Fengyun-1C	2007	2,841	Intentional collision
Cosmos 2251	2009	1,276	Accidental collision
STEP 2 rocket body	1996	713	Accidental explosion
Iridium 33	2009	521	Accidental collision
Cosmos 2421	2008	509	Unknown
SPOT 1 rocket body	1986	492	Accidental explosion
OV 2-1/LCS 2 rocket body	1965	473	Accidental explosion
Nimbus 4 rocket body	1970	374	Accidental explosion
TES rocket body	2001	370	Accidental explosion
CBERS 1 rocket body	2000	343	Accidental explosion

Of the thousands of rocket launches and space missions since Sputnik, only 10 account for about one-third of all cataloged Earth-orbiting debris objects (as of 2010). And 6 of the 10 occurred in the decade between 2000 and 2010. Table 1.1 shows the "top 10" list in which satellite providers don't want to appear.

For those who do not easily relate to graphs, there is another way to illustrate the magnitude of the debris growth. Shown in Figures 1.7–1.9 are artist concepts

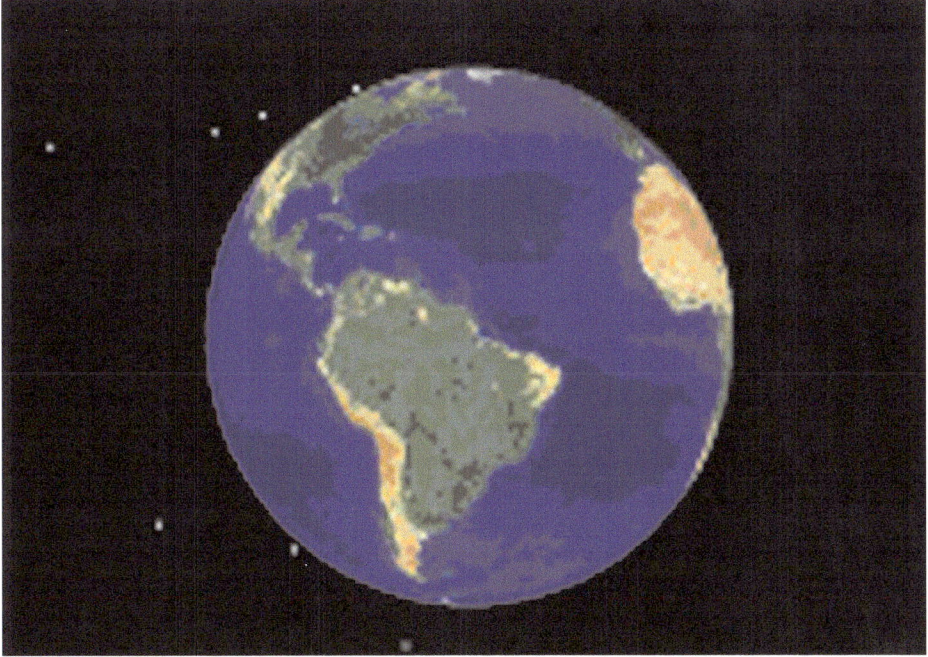

Figure 1.7 In 1960, there were only a few debris objects orbiting the Earth. (Image courtesy of NASA)

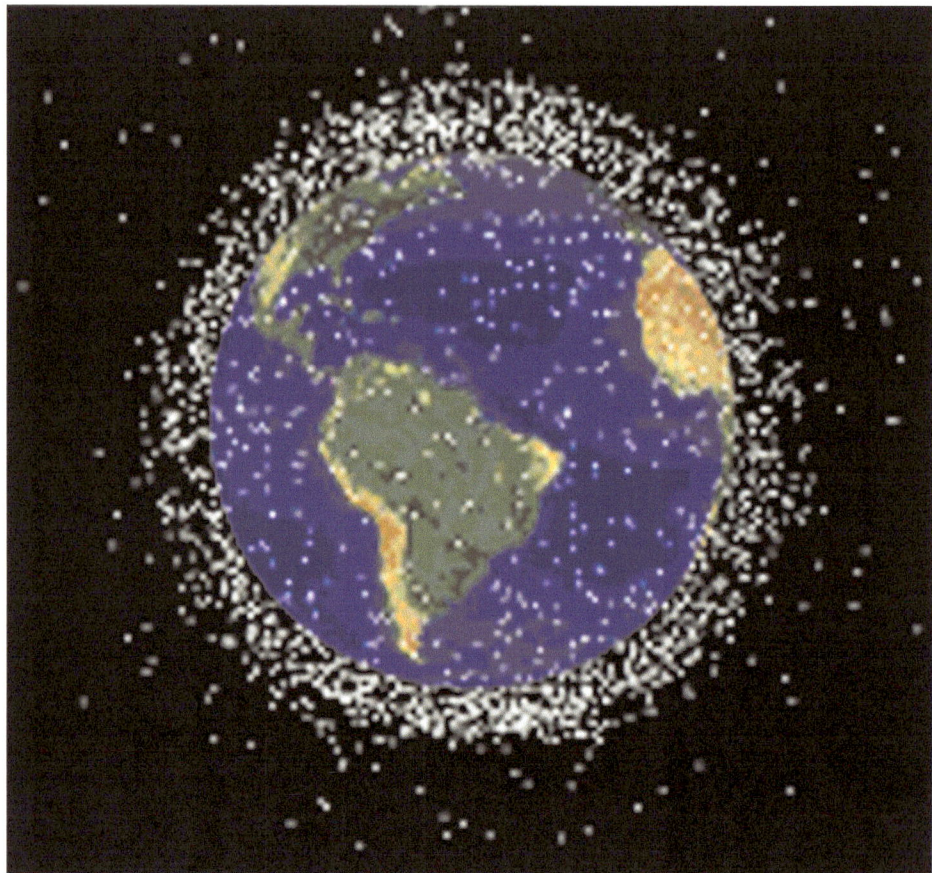

Figure 1.8 By 1980, Earth-orbiting debris objects numbered in the thousands. (Image courtesy of NASA)

showing growth in the amount of debris circling the Earth since the beginning of the Space Age. What matters is the number of debris objects shown; the sizes of the debris dots relative to each other and to the Earth are not to scale. (If they were, you wouldn't be able to see the debris at all. The Earth is 12,756 km in diameter while much of the junk orbiting the Earth is less than 2 cm in diameter!)

I've made the case that we have allowed our creation of space debris to get out of control. But what is the real risk? The total volume of space around the Earth is enormous and, even with so many pieces of junk, the probability of getting hit should still be relatively low. And, as of today, the problem is manageable. New satellites launch into space and the probability of any getting hit by a piece of junk is still rather low. Unfortunately, if the growth rate of debris production continues on its current path, that will not always be the case.

In 1978, NASA orbital debris expert Don Kessler postulated that there may be a

Figure 1.9 By 2010, there were over 1.5 million pieces of debris orbiting the Earth. (Image courtesy of NASA)

tipping point that would ultimately lead to a runaway growth in the orbital debris population in a manner reminiscent of a nuclear chain reaction. The tipping point would be caused when there are enough debris particles in orbit colliding with otherwise intact spacecraft to create yet more debris particles, causing yet more collisions, until the population of debris in orbit would be so large as to create an unacceptably high probability of any new spacecraft being hit by junk early in its lifetime [9]. This space disaster scenario has come to be known as the Kessler Syndrome. (I've always wondered how people feel when a disease or disaster scenario is named after them. I had the chance to meet Don recently and his reply was "I never really thought about it that way". Personally, I would not want to have my name immortalized by its being associated with a scenario that might deny the human race the use of LEO!)

So there you have it. Simply put, we're trashing space. And, unless we do something soon to reverse the process, the space revolution may be self-limiting.

References

[1] Johnson, N.; Liou, J. A Sensitivity Study of the Effectiveness of Active Debris Removal in LEO. *Acta Astronautica*, **64**, 236–243 (2009).
[2] Green, C.M. *Vanguard: A History*. SP-4202, NASA, Washington, DC (1970).
[3] History of On-Orbit Satellite Fragmentations, 14th edn. NASA/TM-2008-214779.
[4] Young, K. Rocket Explosion Creates Dangerous Space Junk. *NewScientist.com* (22 February 2007).
[5] Anselmo, L.; Pardini, C. Analysis of the Consequences in Low Earth Orbit of the Collision between Cosmos 2251 and Iridium 33. 21st International Symposium on Spaceflight Dynamics, Toulouse, France, 2009.
[6] *NASA Orbital Debris Quarterly News*, Volume 12, Issue 1, January 2008.
[7] NASA. USA Space Debris Environment and Operations Update. Presentation to the 47th Session of the Scientific and Technical Subcommittee for the Committee on the Peaceful Uses of Outer Space, United Nations, February 2010.
[8] *NASA Orbital Debris Quarterly News*, Volume 14, Issue 3, 2010.
[9] Kessler, D.J.; Cour-Palais, B.G. Collision Frequency of Artificial Satellites: The Creation of a Debris Belt. *Journal of Geophysical Research*, **83**(A6) (1978).

2 Space War

In Chapter 1, we saw that even a small piece of debris, perhaps something as small as a baseball, traveling at over 8 km/sec can damage or destroy a satellite. It's bad enough that our willful neglect of the space environment over the last 50 years has placed our valuable space assets in harm's way; we now have to consider that someone might intentionally trash the space around the Earth in order to destroy our satellites there.

Since the earliest days of the Space Age, war in space has been contemplated and planned for by the United States and the former Soviet Union. With the rapid advancements in space technology during the Cold War era, both the United States and the Soviet Union feared that the other would gain a strategic military advantage in space that would lead to a potential nuclear first strike, thus turning the Cold War into a hot one. Both countries also realized that regular space satellite over flights of their territories would make their ground forces almost impossible to conceal. As nuclear arsenals grew, so did the concern that one nation or the other would place nuclear weapons in orbit – ready to rain down upon the other's military and industrial complexes without any advance warning. To counter this threat, anti-satellite weapons and strategies were born.

One of my favorite pieces of science fiction art portrays the very realistic fears of the era. The image shows an orbiting space platform containing tens of missiles aimed downward towards the Earth. On the Earth below, the ominous glow of multiple nuclear bomb-induced mushroom clouds rise upward into the upper atmosphere. The imagery is compelling: whoever controls the "high ground" of space would control the Earth. Fortunately, this particular scenario never became reality. As missile technology advanced, the need to place nuclear weapons in space never ended up making strategic sense and the overt militarization of space, the placement of nuclear weapons there, never became a reality.

But that didn't stop military planners from trying to figure out how to destroy an adversary's spy satellites during a crisis. Losing our Global Positioning System (GPS) network, reconnaissance satellites, communication networks, and weather forecasting ability in one fell swoop is too tempting a target for a potential adversary to ignore. The consequences of losing these spacecraft would be devastating and will be described in more detail in later chapters. For now, it is sufficient to say that, without our space satellites, our military would be operating blind and without the command-and-control functions upon which modern-day armed forces depend. But how could this happen? Is it possible to destroy or significantly degrade our network of space satellites as an act of war?

Yes, is it possible and, unfortunately, it would not be too terribly difficult.

Nuclear Weapons

What "better" idea born in the Cold War could there be than the destruction of space satellites by nuclear weapons? If an atomic bomb can destroy a city, surely it can destroy satellites in space. Unfortunately, exploding a bomb in space can do much more than simply damage a spacecraft or two.

In 1958, the United States detonated two nuclear weapons in the upper atmosphere just short of the 100-km altitude now thought of as a border between our atmosphere and outer space. The first, part of Operation Hardtack, successfully exploded at an altitude of 76 km. The bomb injected a significant amount of fission debris into the ionosphere and disrupted radio communication throughout the South Pacific. In addition to the disruption of the ionosphere, the bomb also produced an electromagnetic pulse (EMP). An EMP is a burst of electromagnetic radiation that propagates for many hundreds of kilometers, depending upon where it is detonated, causing damaging current and voltage surges in unshielded devices containing electrical circuits [1].

It wasn't until a subsequent nuclear test in space, the Starfish Prime test of 1962, that the wide-ranging effects of EMP gained appreciation among scientists

Figure 2.1 The Ivy Mike atmospheric nuclear test. (Image courtesy of the US Department of Energy)

and war planners. In July 1962, the United States detonated a >1-megaton bomb in space 400 km above the Earth. The space blast caused EMP-induced damage in Hawaii, over 1,400 km away from the detonation point, by knocking out street lights, setting off burglar alarms, and damaging other electronic equipment. The Starfish Prime explosion was visible from the ground and a similar test, Ivy Mike, can be seen in Figure 2.1. The Soviet Union also tested nuclear bombs in space, producing similar results.

Nuclear bombs detonated in space can destroy satellites in at least two separate ways. The first is the EMP. Traveling at the speed of light, an EMP will sweep across any satellite in view of the bomb. This will affect a lot of satellites – space is large, but there isn't much to obstruct the radiation from the explosion and its intensity will only decrease with distance from the burst. Unfortunately, many hundreds of satellites may be in the line of sight of a space nuclear explosion and close enough to feel its EMP effects.

As the X-ray and gamma-ray radiation impacts the satellite, it induces very high voltages in its electrical conductors, causing them to fail. Systems can be designed to survive this rapid pulse of radiation, but it is expensive and increases the weight of the satellite [2]. Military systems have this sort of shielding, which should be sufficient to protect them from a burst unless it occurs very close to the satellite itself. Unfortunately, most civilian satellites are not similarly radiation hardened and are subsequently at much higher risk.

The second way a satellite can be damaged by nuclear detonations in space is by the "pumping" of the Earth's radiation belts.

Consisting mostly of protons and electrons, these radiation belts are regions where high-energy particles are captured by the Earth's strong magnetic field. Charged particles like electrons and protons experience a force when they move through a magnetic field. In this case, the force serves to keep them trapped in specific regions surrounding the Earth. In 1958, James Van Allen discovered an intense low-altitude radiation belt, which was subsequently named in his honor: the Van Allen belt [3].

Satellites pass through these radiation belts all the time and are designed to survive their encounters with the typical radiation doses induced by their passage through them. It is believed that the charged particles contained in these belts, of which there are two primary ones, originated within the solar wind. (The solar wind is nothing more than a stream of charged particles, including electrons and protons, streaming constantly into space from the Sun.) The Earth's radiation belts are illustrated in Figure 2.2.

Usually, spacecraft are launched into orbits that will allow them to pass through the Van Allen belts quickly and not traverse them very often. Even the best spacecraft shielding is limited in the extent to which it can protect against this type of radiation. The design approach is therefore one of avoidance rather than survival within the belts.

A nuclear detonation releases gamma rays (very high-energy electromagnetic waves), which interact with the atmosphere by stripping atoms of their electrons, creating positively and negatively charged ions. This, coupled with the radioactive

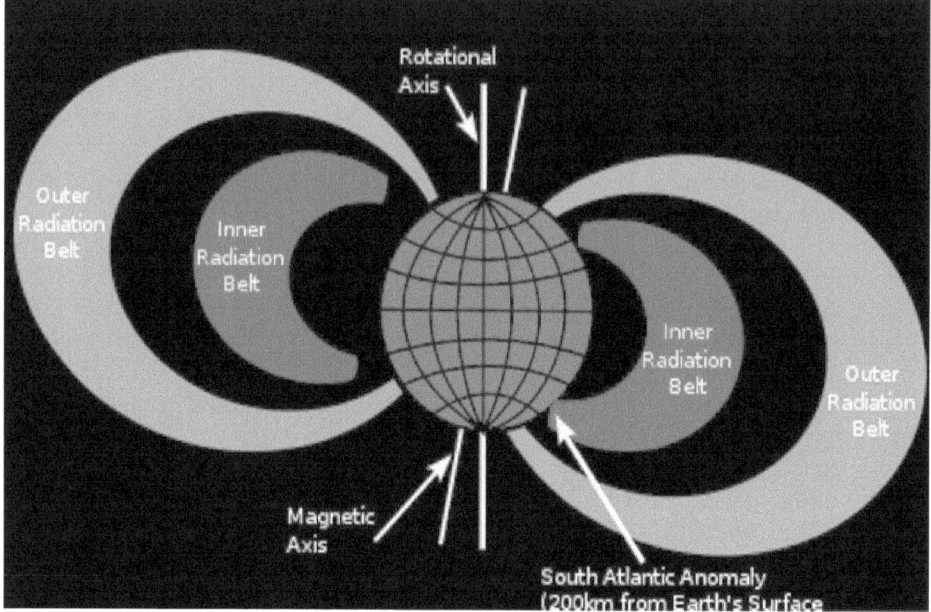

Figure 2.2 The Earth's Van Allen radiation belts extend outward from the Earth and comprise trapped protons and electrons. (Image courtesy of NASA)

decay of the bomb's fragments, pumps the Van Allen belts with energetic electrons. Satellites then fly through these pumped/enhanced radiation belts and experience radiation doses far above their design limits, causing the progressive failure of critical spacecraft systems that quickly exceed their lifetime radiation dose limits [4].

Following multiple nuclear weapon explosions in space, there will be hundreds of dead and dying spacecraft circling the Earth on their way to becoming orbital debris.

Electric Fire

To destroy a satellite, an adversary doesn't even have to go to space. Ground-based weapon systems are under development that could strike satellites as they pass overhead. Some of these weapons wouldn't necessarily destroy a satellite immediately, perhaps only causing gradual degradation of the spacecraft's capabilities or causing some of its subsystems to fail with each orbital pass, making it more difficult to determine which country or individuals within a given geographic area are responsible.

High-energy laser weapons resembling in appearance their science-fictional counterparts can degrade or destroy virtually any optical system on board a spacecraft (Figure 2.3). And today's satellites use many optical systems, including

Figure 2.3 NASA GSFC's Laser Ranging Facility was used to send a laser beam towards the Lunar Reconnaissance Orbiter in orbit around the Moon to obtain extremely accurate distance information. A much more powerful system could be used to blind or damage satellites in Earth orbit. It must be noted that NASA does not develop weapons systems! (Image courtesy of NASA)

telescopes, optical communication arrays, Sun sensors (to monitor the location of the Sun in order to keep the spacecraft pointed in the right direction during flight), and solar power generating arrays. Solid-state laser technology has advanced dramatically in the last 20 years, making the prospect of high-power, mobile, and very powerful lasers capable of damaging spacecraft flying hundreds of miles overhead a very real possibility in the near future.

Space-based neutral particle beams, similar to the beams generated in high-energy research facilities like the European Organization for Nuclear Research, commonly known as CERN, can propagate through the vacuum of space and scramble the electronics of spacecraft, causing either sporadic or total failure of on-board systems. During the 1980s, the United States was publicly developing such weapons as part of the Star Wars defense system. Particle beams, consisting of neutral hydrogen and its isotopes, cannot pass through the Earth's atmosphere – collisions between the atoms in the beam and the atmosphere would quickly scatter the beam, rendering it ineffective. But a beam generated in space from aboard a large, orbiting platform could quite easily target other satellites many hundreds or thousands of kilometers away.

Perhaps the simplest "ray gun" is some sort of high-power, highly directional radio or microwave beam that could jam radio communications with a satellite at relatively low power or burn out its electronics at high power. Sometimes called "electric fire", the technology behind this type of weapon is not that much different from what is used in modern radar systems.

The good news is that the threat posed by these futuristic anti-satellite weapons has been known for quite some time, and it is highly likely that military satellites have some sort of countermeasures they can employ to defend themselves. Once the threat is known, a satellite might perform evasive maneuvers to get out of harm's way or deploy chaff to reduce the effectiveness of a laser or particle beam. Just as a satellite can be hardened against the effects of nuclear radiation, it might also be hardened against the effects of a microwave or laser beam.

The bad news is that civilian spacecraft, like weather, communications, and science satellites, are likely totally unshielded and unprotected against these threats.

Hackers

As we've learned during the last few years, hackers are much more sophisticated than ever before and are constantly challenging the integrity of commercial and government computer networks. Apparently, they are also targeting our satellites. As reported in the 2011 Report to Congress of the US–China Economic and Security Review Commission:

> "Satellites from several U.S. government space programs utilize
> commercially operated satellite ground stations outside the United

States, some of which rely on the public Internet for 'data access and file transfers,' according to a 2008 National Aeronautics and Space Administration quarterly report. The use of the Internet to perform certain communications functions presents potential opportunities for malicious actors to gain access to restricted networks.

Notably, at least two U.S. government satellites have each experienced at least two separate instances of interference apparently consistent with cyber activities against their command and control systems:

- On October 20, 2007, Landsat-7, a U.S. earth observation satellite jointly managed by the National Aeronautics and Space Administration and the U.S. Geological Survey, experienced 12 or more minutes of interference. This interference was only discovered following a similar event in July 2008 (see below).
- On June 20, 2008, Terra EOS [Earth observation system] AM–1, a National Aeronautics and Space Administration-managed program for earth observation, experienced two or more minutes of interference. The responsible party achieved all steps required to command the satellite but did not issue commands.
- On July 23, 2008, Landsat-7 experienced 12 or more minutes of interference. The responsible party did not achieve all steps required to command the satellite.
- On October 22, 2008, Terra EOS AM–1 experienced nine or more minutes of interference. The responsible party achieved all steps required to command the satellite but did not issue commands.

The National Aeronautics and Space Administration confirmed two suspicious events related to the Terra EOS satellite in 2008 and the U.S. Geological Survey confirmed two anomalous events related to the Landsat-7 satellite in 2007 and 2008.

If executed successfully, such interference has the potential to pose numerous threats, particularly if achieved against satellites with more sensitive functions. For example, access to a satellite's controls could allow an attacker to damage or destroy the satellite. The attacker could also deny or degrade as well as forge or otherwise manipulate the satellite's transmission. A high level of access could reveal the satellite's capabilities or information, such as imagery, gained through its sensors. Opportunities may also exist to reconnoiter or compromise other terrestrial or space based networks used by the satellite."

www.uscc.gov/annual_report/2011/annual_report_full_11.pdf

It is important to note that, although the report quoted above implies that the interference with our satellites was caused by the Chinese due to its inclusion in a report about China, the report doesn't explicitly name them as the perpetrators. It is possible that some third, as-yet-unknown party is testing a capability to take control of some of our satellite infrastructure.

Space Rocks

Orbital debris was discussed in the previous chapter, as was the increase in the orbital debris population resulting from the Chinese testing of an anti-satellite weapon in 2007. The deliberate destruction of the Chinese Fengyun-1C polar-orbiting weather satellite created a cloud of debris that will threaten satellites for centuries or longer. This increased threat resulted from the willful destruction of only one satellite. What if a country decides to attack multiple satellites, with the goal being not the immediate destruction of any particular spacecraft, but rather the creation of a cloud of orbital debris so large that all spacecraft passing through it would have a significant chance of being destroyed?

The message sent by China to the United States and Russia was just that. Their destruction of Fengyun-1C was not simply "Hey, we can destroy a satellite in space!" Rather, it was "Hey, we can destroy all your satellites in space!" What they may not fully appreciate is that doing so would also deny themselves the capabilities provided by satellites in space. Making the space environment inhospitable to satellites would not be in the economic or military self-interest of China as a rising industrial and space power. I believe this because, like the Soviet Union and the United States during the Cold War, China is run by rationale leaders who will not take an obviously self-destructive course of action to achieve a short-term political goal. The same cannot be said of other rising space powers – take Iran, for example.

Iran launched its first Earth-orbiting satellites in the early 2000s. Any country with the capability of launching satellites into space has the inherent capability of building an anti-satellite system and destroying the satellites of another nation. And it doesn't have to be a sophisticated system. It can be as simple as intentionally colliding your spacecraft with that of another nation, destroying it and creating thousands of pieces of additional space debris.

Iran is a theocracy. Its leaders have repeatedly called for the destruction of both Israel and the West. The country has been implicated in multiple acts of international terrorism and it is not difficult to imagine its leadership launching non-nuclear attacks on the United States and developed world's satellite infrastructure to wreak as much damage as possible. With rising tensions in that region of the world, should Iran feel threatened or be attacked, it is easy to see how they might to decide to retaliate with the destruction of as many satellites as possible – dramatically increasing the debris risk for all countries of the world in the process.

In addition to using anti-satellite weapons like China, Russia, and the United States have in their respective arsenals, any country that can access space could use a low-tech version of an ASAT weapon to take out a large number of satellites anywhere in space. And it can be as simple as the pebbles on a beach.

Recall the Chapter 1 discussion of the damage that can be caused by a small rock traveling at ~8 km/sec. Now imagine that a rocket containing tons of small rocks is launched into space with the intention of placing those rocks into orbits used by military, communication, or navigation satellites. The immediate effect would be like a shotgun blast, with individual pellets containing more than 100 times the kinetic energy of a 9-mm round. And a single rocket could loft more pebbles into space in one launch than all the debris objects currently in orbit – hundreds of thousands of rocks per rocket launch. It would take time for the rocks to disperse and become a serious threat to those satellites not in the immediate vicinity at the time, but they would spread out, eventually forming a spherical shell around the Earth.

To understand why the rocks, and any debris cloud for that matter, spread out from their initial orbit to encircle the Earth in a roughly spherical shell, we must remember several things:

- The Earth is not a perfect sphere and its gravity field is not uniform. The Earth is technically an oblate spheroid, which means it is roughly spherical with a bulge around the equator. Since the gravity field is determined by the planet's mass distribution, the bulge around the equator warps the field and keeps it from being uniform.
- The Earth is orbited by a large moon with its own gravitational field. Objects orbiting the Earth have their motion affected by the Earth's gravity and that of the Moon. For the same reason that there are tides, objects orbiting the Earth will be subject to the Moon's gravitational tugs.
- Each piece of debris, or in this case each rock, will have its own three-dimensional velocity that is different from those around it. This is especially true if the debris object is affected by an explosion – such as that which might be used to disperse thousands of rocks ejected from a rocket into space. Even in a uniform gravity field, these small differences, over time, will cause each piece of debris to move away from its neighbors.

Taken together, the result will look like Figure 2.4. Just after the event, the cloud of objects remains in a fairly tight orbit circling the Earth in roughly the same orbit as the rocket that lofted them. A few months later, the orbits of individual pebbles will have drifted sufficiently to cover a wider band. A year later, the cloud will have dispersed to near-global coverage and, within a few years, the cloud of objects will intersect nearly every Earth-orbital slot within a fairly well-bounded orbital altitude regime.

Satellite Self-Detonation

The willful destruction of our satellites need not be as spectacular as nuclear war. After all, if hundreds of nuclear bombs are exploding in space, there will likely be many more going off on the ground at the same time and the loss of our satellite capabilities will be only one item on a very long list of capabilities lost – too far

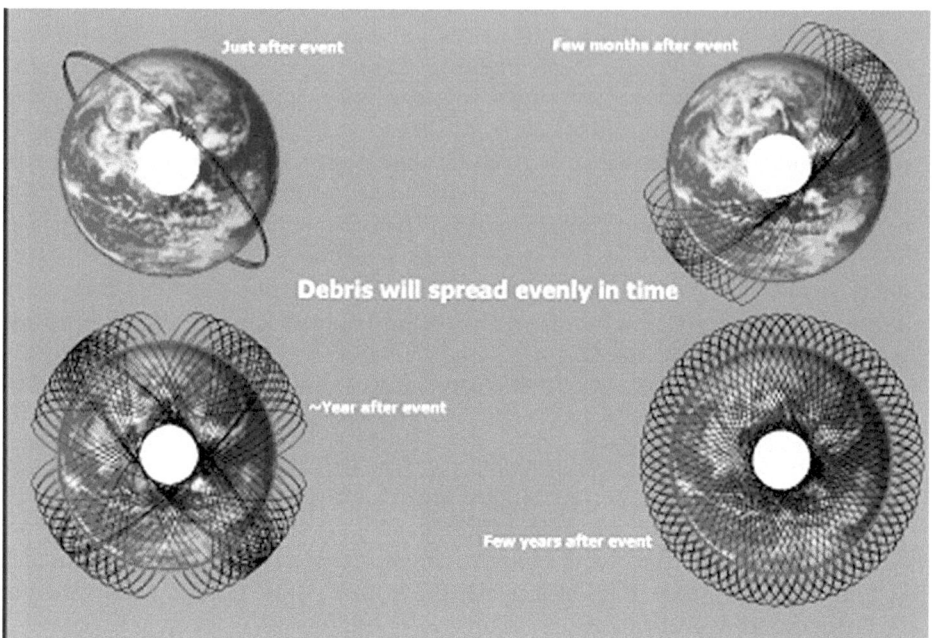

Figure 2.4 An orbital debris cloud created in one orbit will eventually spread out and encircle the globe. (Image courtesy of NASA)

down on the list, I should suppose, for the average person who might still be alive to worry about.

It also need not be as overtly hostile as a country using its anti-satellite weapons to target specific satellites in a non-nuclear battle – say somewhere in the Middle East or in the South China Sea. Using even one anti-satellite weapon in an otherwise "limited" regional conflict might be perceived as an escalation towards a nuclear war, thereby constraining a country from taking that step lightly.

No, and as history suggests, it is possible a country might willfully destroy its own spacecraft to avoid having them captured by a potential adversary. This willful self-destruction would lead to the creation of orbital debris, contributing to the possibility of the Kessler Syndrome actually occurring. Why is this outlandish scenario even considered?

The recently retired US Space Shuttle was a fantastic machine capable of not only carrying a crew to and from space, but also of:

- remaining on orbit for weeks using pressurized modules in the cargo bay for scientific research; NASA flew a series of Space Lab missions doing just that;
- assembling the International Space Station; in fact, dozens of Shuttle flights carried individual pieces of the station to orbit and the crew used the orbiter as their home while assembling it; a key capability for facilitating this was the Shuttle's robotic arm;

- repairing and releasing repaired satellites; the most famous examples of satellite repair were the Hubble Space Telescope repair and refurbishing missions; using the robotic arm, astronauts captured the Hubble and installed multiple new instruments and replacement parts in the course of five separate missions (Figure 2.5).

If you can capture and repair your own satellite, what's to stop you from capturing one that doesn't belong to you? It was perhaps this fear that motivated the Soviet Union to equip an entire generation of satellites with self-destruct capabilities as a deterrent. After all, would we want to capture a satellite and have it explode in the Shuttle's cargo bay?

Figure 2.5 The Space Shuttle captured and serviced the Hubble Space Telescope using its robotic arm in 1993 on STS-61. (Image courtesy of NASA)

The Soviet Union, and later Russia, is thought to have intentionally exploded over 50 spacecraft. Were some of these deliberate explosions meant to send a message of "don't mess with our satellites?" It is entirely possible that this was the case. (Again, each explosion contributed to the total population of orbital debris.)

Our satellites form a very fragile infrastructure upon which so much of our well-being and economic livelihood are based. Is it a good idea to have our survival as a civilization based on something that can be so easily destroyed? As a space advocate, I say "yes!" As a cautious and somewhat analytical person, taking into account the costs and benefits versus the risks, I am not so sure. In any case, we should be developing satellites that are harder to destroy and some sort of backup capability that can rapidly be put in place should the worst happen. It would be almost criminal not to do so.

References

[1] Hoerlin, H. United States High-Altitude Test Experiences: A Review Emphasizing the Impact on the Environment. Report LA-6405, Los Alamos Scientific Laboratory (October 1976).
[2] Stassinopoulos, E.G. The Space Radiation Environment for Electronics. *Proceedings of the IEEE*, **76**(11) (1988).
[3] Chen, Y.; Reeves, G.; Friedel, R. The Energization of Relativistic Electrons in the Outer Van Allen Radiation Belt. *Nature Physics*, **3** (2007).
[4] Rodger, C.J., et al. The Atmospheric Implications of Radiation Belt Remediation. *Annales Geophysicae*, **24**, 2025–2041 (2006).

3 Solar Storms

We've only been sending craft into space for a little over 50 years. The Earth is 4.5 billion years old. My point? We do not have enough data to know whether the space environment our satellites have experienced since the dawn of the Space Age is normal, a relatively quiet period, or if we're in the middle of a particularly active period for the Sun and its effects on our near-space environment. There just aren't enough data.

Why is this important? To answer this question, we need to discuss the Sun and its impact on the Earth and the space near our planet. In addition to providing us with light and heat, the Sun's nuclear furnace sends into space a nearly constant flow of other forms of radiation that are not so benign.

Take, for example, the solar wind. The solar wind is a stream of charged particles produced by the Sun and sent into the Solar System. Composed mostly of protons and electrons, the solar wind extends outward from the Sun and its effects can be felt as far away as the outer Solar System – well beyond the orbit of Pluto. While not constant, the solar wind is fairly well understood and the radiation contained within it can be accounted for in spacecraft design [1]. (See also Chapter 2 for the effects of radiation, including the solar wind, on spacecraft.)

Unfortunately, the radiation output of the Sun is not limited to the solar wind. The Sun is, in fact, a slightly variable star with a well-defined 11-year cycle. This is most visibly indicated by the frequency of the occurrence of sunspots. Sunspots are regions where the Sun develops intense local magnetic activity, producing increased X-rays and particle radiation (such as protons). The magnetic activity reduces the amount of visible light emitted by the Sun, making the Sun slightly cooler; hence, sunspots appear black on the surface of the Sun (Figure 3.1). For reasons that are not yet well understood, the Sun's output increases and decreases over each cycle and then repeats.

The intense magnetic activity associated with large sunspots can result in solar storms. A solar storm is a period of intense activity that can result in the generation of an X-ray flare, a coronal mass ejection, or usually both. An X-ray flare is an intense pulse of X-rays, anywhere from one minute to a couple of hours in duration. A coronal mass ejection, abbreviated CME, is an ejected ball of very hot gases coupled by magnetic fields. A CME can be many times larger than the Earth. While the gas is not dense enough to pose a thermal hazard, the particles in the gas are moving rapidly enough to cause significant radiation damage to affected equipment and personnel. A CME can expel a billion tons of matter into space. The frequency of solar storms is another characteristic of the solar cycle, with more storms coming during years near the sunspot maximum.

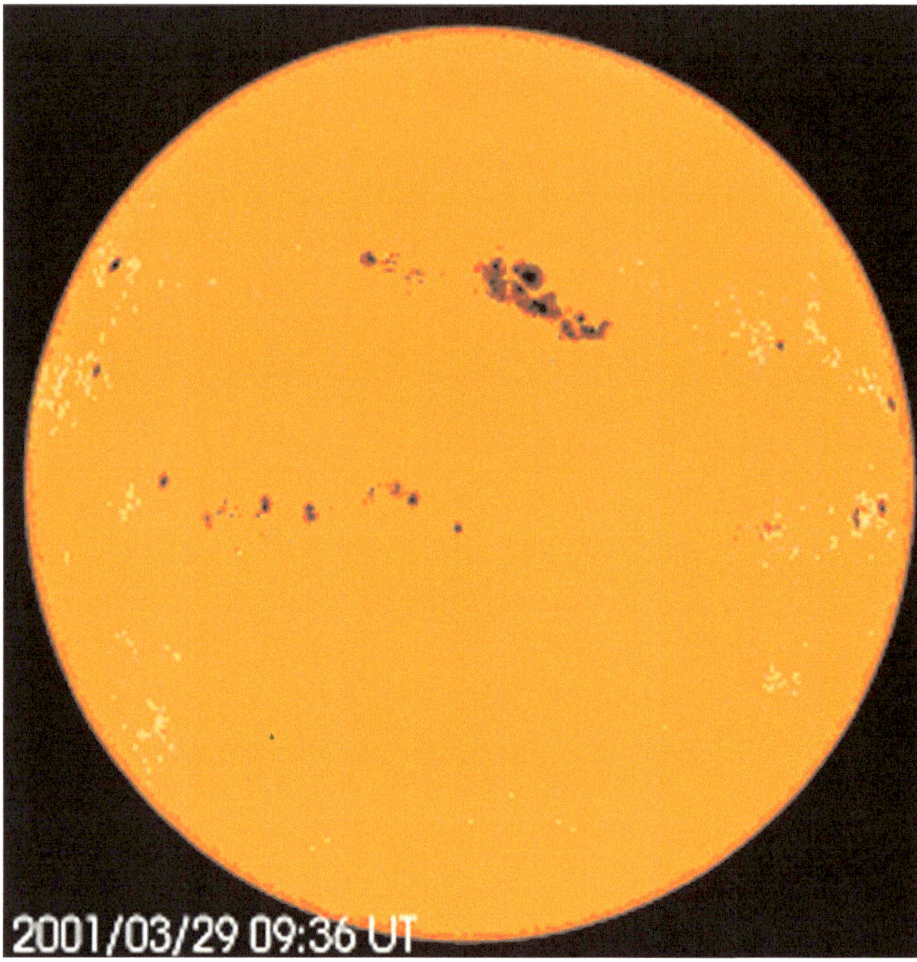

Figure 3.1 The Sun with a visible array of sunspots. (Image courtesy of NASA)

Intense pulses of fast-moving protons can be produced with a solar X-ray flare. The Sun experiences such a solar proton event several times each year. They can happen at any time, although their frequency of occurrence varies with sunspot number over the solar cycle.

A solar storm is an Earth-sized (or larger) burst of high-energy radiation (protons, electrons, and alpha particles) thrown into space by the Sun. Hurtling outward from the Sun at 500–1,000 km/sec, these storms often cross the Earth's orbit, engulfing the planet in their fury.

Figure 3.2 shows a coronal mass ejection (left) as seen from NASA's Solar Terrestrial Relations Observatory (STEREO) spacecraft. The image shows only the outermost layer of the Sun's atmosphere. An occulting disk covers the rest of the Sun and the white circle in the middle of the disk represents the relative

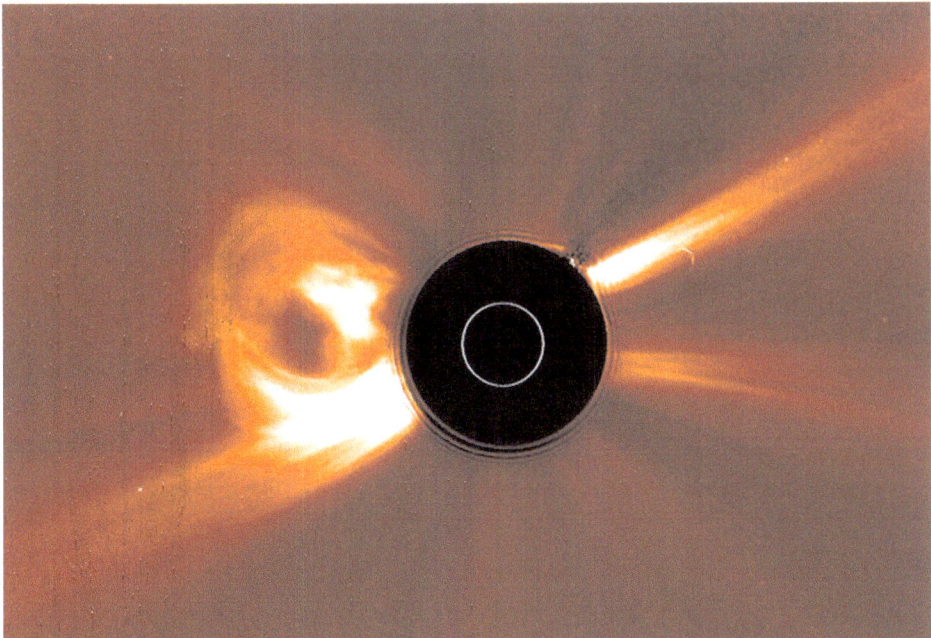

Figure 3.2 A mass of charged particles (left) loops outward from the Sun. (Image courtesy of NASA)

location and size of the Sun's surface. For reference, the Sun's diameter is about 100 times that of the Earth. This means that about 100 Earths, sitting side by side, would stretch across the face of the Sun.

When a storm reaches the Earth, many of us living on the ground don't even notice its arrival. First of all, much of the charged-particle radiation (protons and electrons) is either deflected or absorbed by the Earth's magnetosphere. Thanks to the Earth's very strong magnetic field, a magnetic bubble containing these charged particles surrounds us, trapping much of the radiation before it ever enters the atmosphere. Figure 3.3 shows a not-to-scale interaction of a solar storm with the Earth's magnetosphere, illustrating the shielding effect it provides.

The radiation that is too energetic to be trapped by the field enters the Earth's dense atmosphere and is absorbed long before it reaches the surface of the planet. This process can have a direct effect on people on the Earth. Particularly if you live near one of the Earth's magnetic poles, you may indirectly notice a solar storm by an increase in the strength of the aurora. The trapped particles spiral along the Earth's magnetic field, enter the upper atmosphere at the North and South Poles, and collide with atmospheric atoms, causing the sky to glow. In extreme cases, the amount of radiation being channeled into the magnetic poles can present a potential health hazard, causing "over the pole" commercial air flights to be re-routed. In addition, the radiation causes electrical charge build-up

Figure 3.3 The Earth's magnetosphere acts as a shield, protecting the planet from deadly radiation coming from the Sun. (Image courtesy of NASA)

in the upper atmosphere. This can affect high-frequency and even very-high-frequency (short-wave) radio transmissions and, in extreme cases, can couple to electrical transmission lines on the ground, causing power surges resulting in power outages.

In space, the story is different. Satellites are protected against solar storms in the same way as they are shielded against the solar wind – by adding extra mass to absorb the radiation. There is a problem with this approach, however. Simply adding mass may actually make the problem worse before making it better [2].

When an energetic solar wind atom hits any other collection of atoms or matter, it has a probability of interacting with that matter, losing energy (which is what we want), and creating a cascade of secondary particle radiation (which is not what we want). This is due to the fact that any particle of matter has a cross-section of interaction that determines its probability of colliding with another atom when it is passing through at some velocity. The incoming particle collides with an atom or atoms in the matter through which it is traveling, producing new particles with lesser energy. Each of these new particles then interacts with the matter, producing yet more (and lower-energy) new particles. Eventually,

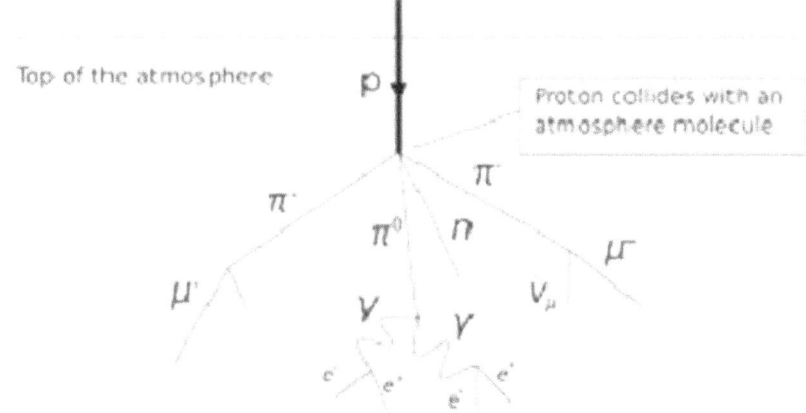

Figure 3.4 A particle cascade in the atmosphere caused by an incoming solar wind proton can produce many secondary particles, each capable of producing yet more particles in subsequent interactions. Similar particle cascades occur internally to satellites exposed to solar proton radiation. (Image courtesy of Frank McNally)

given enough matter, the energy of the secondarily produced particles, though there may now be millions of them, is low enough that they are simply stopped and absorbed. It is this absorption that causes damage to electronics and our spacecraft. Figure 3.4 shows what a particle cascade can look like schematically.

Cosmic rays, which are very-high-energy particles produced in the very distant universe, may have sufficient energy to simply pass through our spacecraft without being absorbed and doing minimal damage. At first thought, this may seem counterintuitive. Those coming from the Sun are typically much lower in energy than these galactic cosmic rays, and that will cause damage to the spacecraft and its electronics. A faster-moving atom seems like it should do more damage, not less. But, in fact, a high-energy atom will simply pass through and not have sufficient "interaction time" to deposit enough energy to cause damage. If we apply some shielding to stop the lowest-energy cosmic rays before they can damage the spacecraft electronics, we may inadvertently create many, many more damaging particles due to their interaction with the higher-energy cosmic rays.

Designing a spacecraft to survive the radiation accompanying a coronal mass ejection is a tricky thing. You need to have enough mass to stop the very damaging low-to-mid-energy radiation without having enough to create a deadly cascade of secondary radiation from the slightly more energetic ones. Designing to solve one problem may lead to a very different, but related, one that is much more difficult to resolve.

The strength of a solar storm can exceed the design limits of a spacecraft, causing its damage or its destruction. Fortunately, this has rarely happened.

But will this always be the case? Recall the question asked at the beginning of

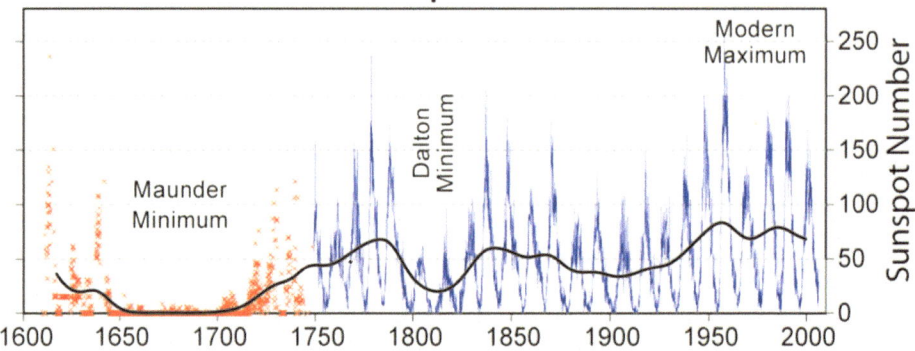

Figure 3.5 The solar cycle is observed to have continued in roughly its present form since the early 1700s. Prior to that, the Sun was very quiet in a time known as the Maunder Minimum. Another, smaller, minimum occurred in the early 1800s – the Dalton Minimum. (Image courtesy of Robert A. Rohde)

this chapter. Will the Sun always produce levels of radiation that we can manage? We've only been in space for a little over 50 years and we've only been observing the solar cycle for a few hundred years. Prior to the Space Age, the primary way of gauging solar activity was to count the number of sunspots seen on the Sun. Figure 3.5 shows solar activity over the last 400 years. The frequent peaks and valleys correspond to the approximate 11-year solar cycle and the continuous dark line running through the peaks indicates the running annual average.

It is interesting to note that there was a period from the early 1600s to the mid-1700s during which sunspots and solar storms almost disappeared. Will this happen again?

In 2008, there were some who thought so. Figure 3.6 was taken on 27 September 2008 by the space-borne Solar and Heliospheric Observatory (SOHO). It shows the solar disk with no sunspots. 2008 now has the distinction of being the "blankest year" (with regard to sunspots) of the Space Age [3]. In fact, as of the time this photograph was taken, there had been no sunspots for 200 days that year. To find a year with that many blank days, you would have to go back to 1954 – three years before the first satellite was launched into space.

The sunspots eventually returned and the solar cycle continued.

If we can today experience periods of extremely low, and previously unrecorded minimal, solar activity, might we enter a period with peaks much larger than that experienced in the last few centuries, producing storms of greater intensity – greater than our spacecraft are designed to handle?

Consider the solar storm of 1859, which was the largest solar storm ever observed. The solar flare associated with the storm was observed visibly by London astronomer Richard C. Carrington, who was recording the associated sunspot at the time of the storm (Figure 3.7). A magnetic inflection was observed

2008/09/27 17:36

Figure 3.6 2008 marked the minimum of the recent solar cycle. This image, taken by the SOHO spacecraft, shows the Sun with no visible sunspots. (Image courtesy of NASA)

at Greenwich Observatory at the same moment (due, as we know now, to the flare X-rays impacting the atmosphere), permitting the astronomers of the time to draw a correlation between the flare and the subsequent aurora and geomagnetic events. The storm was so powerful that the resulting auroras were visible as far south as Cuba, and operation of the telegraph, the highest technology communication system of the time, was severely disrupted [4]. There were reports of failed telegraph systems throughout North America and Europe, with some telegraph stations reportedly catching fire as a result of the currents induced in the wires when the solar storm impacted on the Earth's magneto-sphere and magnetic field.

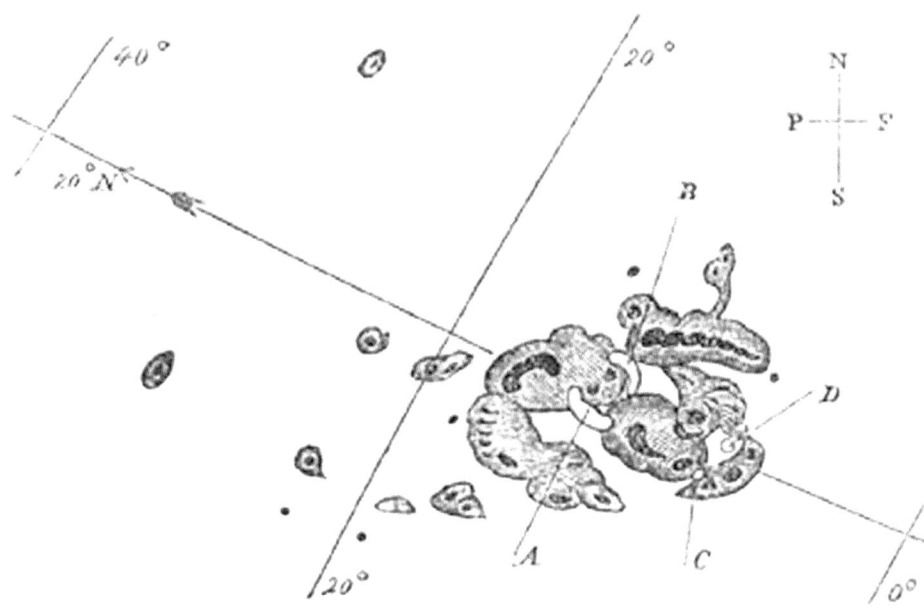

Figure 3.7 Sketch of the sunspot giving rise to the 1 September 1859 solar flare. Richard C. Carrington observed the visual flare rise at points A and B, then spread to points C and D, before the flare dissipated after about five minutes. (Originally published in *Monthly Notices of the Royal Astronomical Society*, Volume 20, page 13, November 1859, copyright expired)

Since the beginning of the Space Age, the strongest solar storms were observed in March 1989 and around Halloween of 2003. With X-ray observatories in space, the strongest observed X-ray flare was the so-called X28 flare in the Halloween Flare series; however, this flare saturated the X-ray detectors of the GOES spacecraft, and some scientists believe that flare was in fact almost twice as strong as they were able to record [5]. These events were associated with significant recorded in-space radiation environments as well as defined power outages on the ground.

Most scientists agree that a repeat of the Carrington Event today could result in over one trillion dollars-worth of damage to terrestrial electrical and communications grids due to the induced currents. However, the concern here is with the impact of such an intense radiation dose – due both to the flare X-rays and to the subsequent solar storm – on space systems. Few non-military space systems are designed to continue operating through a significant radiation event and spacecraft containing human crew would be very much at risk [6]. With warning, most commercial satellites can shut down temporarily to limit damage from a flare. But, with no way to quantify the dose that was associated with the 1859 storm, we cannot assess what levels of radiation we should design to and what level of temporary service interruption we can permit as we become increasingly dependent on real-time communications with space assets.

Will we experience an event like the 1859 storm again? Almost certainly. The only question is when it will occur.

References

[1] Webb, D.; Howard, R. The Solar Cycle Variation of Coronal Mass Ejections and the Solar Wind Mass Flux. *Journal of Geophysical Research*, **99**(A3) (1994).

[2] Shea, M.; Smart, D. Space Weather: The Effects on Operations in Space. *Advances in Space Research*, **22**(1) (1998).

[3] Nandy, D., et al. All Quiet on the Solar Front: Origin and Heliospheric Consequences of the Unusual Minimum of Solar Cycle 23. *Sun and Geosphere*, **7**(1) (2012).

[4] Committee on the Societal and Economic Impacts of Severe Space Weather Events: A Workshop, National Research Council. Severe Space Weather Events: Understanding Societal and Economic Impacts: A Workshop Report. National Academies Press (2008).

[5] Tsurutani, B., et al. The Extreme Halloween 2003 Solar Flares, ICME's, and Resultant Ionospheric Effects: A Review. *Advances in Space Research*, **37**(8) (2006).

[6] Townsend, L.W. The Carrington Event: Possible Doses to Crews in Space from a Comparable Event. *Advances in Space Research*, **38**(2) (2006).

Part 2

If We Were to Lose
Our Satellites . . .

4 The Global Positioning System (Military Uses)

To understand what the loss of America's Global Positioning System (GPS) would mean to the military, we need to first understand what it is and why it is used.

Since 1995, a network of between 24 and 32 American satellites has been orbiting the Earth, providing continuous reliable position and navigation services to users around the world (Figure 4.1). Built to support the needs of

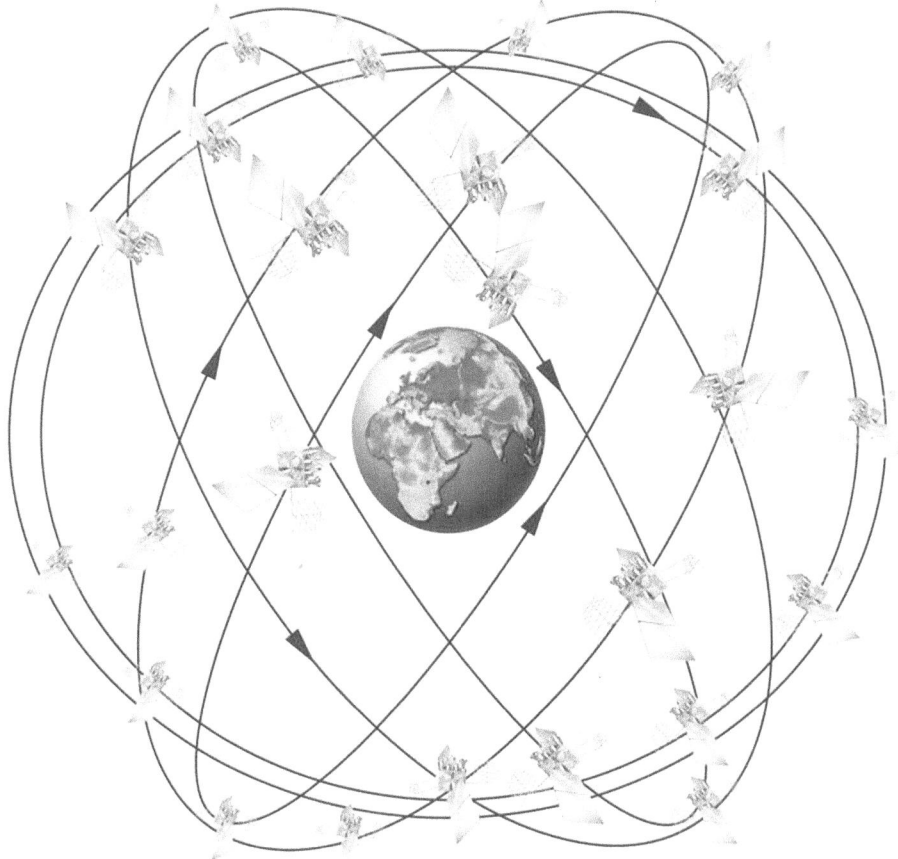

Figure 4.1 The Global Positioning System (GPS) consists of between 24 and 32 satellites, providing continuous location services anywhere on the globe. (Image courtesy of the National Oceanic and Atmospheric Administration)

the US military, the system is now widely and freely used by individuals and commercial companies for navigation, surveying, tracking shipments and commerce, and many other applications that require precise timing and location information.

GPS is used in virtually all US military systems. In fact, Congress mandated that, after the year 2000, any aircraft, ship, or armored vehicle not equipped with GPS will not be funded [1].

In order for it to work and provide precise location information, there must be at least four satellites within "view" of the GPS receiver. In other words, there must be an unobstructed line of sight between the receiver and at least four satellites. Each of the original satellites is placed in an orbit 35,786 km high, thus orbiting the Earth every 12 hr. Each satellite broadcasts its own unique signal and when a receiver picks up signals from multiple satellites, it can use the data from each satellite signal to determine distance and, with multiple signals arriving simultaneously, the receiver can know its position with an accuracy of less than a few meters.

Many consider the Gulf War of 1990–91 to be the world's first war won using space. America and its allies used space reconnaissance, GPS, and GPS-guided munitions to rapidly decimate the Iraqi Army. America's former Cold War adversaries had supplied Iraq with arms and the brief conflict showed that the application of space technology was a critical element in their rapid defeat. Soldiers, tanks, aircraft, and ships used GPS to know where they were with unprecedented accuracy. GPS-guided bombs struck with precision unrivaled in the history of warfare.

The Gulf War is replete with stories of how GPS gave the Allied troops a significant tactical advantage. There were reports of ground and tank forces entering sandstorms, which would have previously made them un-navigable, and maneuvering within them as if the storm were not even occurring. The same benefits accrued to foot soldiers and special operations teams during night-time operations, as they used GPS navigation to find their current location and it helped them get to where they needed to be.

A common problem in war is the tragedy of friendly fire. In the confusion of battle, it is easy to mistake friendly troops for hostile and accidentally target them instead of the bad guys. Today, information about every soldier and weapons system in the battlefield is tracked and fed into a data system called the GPS Truth Data Acquisition, Recording, and Display System (TDARDS).

Now imagine navigating the deserts of Iraq during darkness and losing the GPS signal. Where are you? Where are the various patrols that were nearby? Will the helicopter flying overhead recognize you or paint you as a target? (And, oh yes, that helicopter above you will be flying blind having lost GPS signals itself.) Worst of all, how will you make it to the rendezvous point for extraction when your mission is accomplished if you don't know where you are or how to get there?

On a strategic level, GPS allows nuclear missiles to be targeted with extreme accuracy, allowing the manufacture of smaller warheads, since there is no need to increase a weapon's yield to make up for minor targeting inaccuracies. This is especially needed by submarine-based nuclear missiles, cruise missiles, aircraft, and any non-stationary land-based missiles.

If an adversary were to strike when we're scrambling to operate without GPS, how well would the United States and her allies be able to respond? Consider the fact that many countries have built their military infrastructure around the availability of GPS. Precise munitions mean that you don't need as many bombs or missiles to destroy a specific target – so you don't build as many. GPS-guided munitions are more expensive than "dumb" bombs, like those used in the Second World War, so we cannot afford to build as many of them. The end result is that we don't have as many weapons in our stockpiles as we used to have in the pre-GPS past.

Look at Figure 4.2, showing an Allied bombing raid over Germany from the Second World War. What is most striking about the photograph? *The number of bombs being dropped is enormous.* When you consider that some air raids had over 1,000 bombers, each carrying and releasing loads like these in a little over an hour, you can quickly see that the number of bombs needed to destroy a series of targets in the pre-GPS era was of a scale unimaginable today. The same targets today could probably be destroyed with far fewer aircraft and significantly fewer bombs.

Without GPS, our expensive and now "dumb" bombs become not much more capable than those used in abundance during the Second World War. How many

Figure 4.2. American bombers over Germany during the Second World War. (Image courtesy of the US Air Force)

of these do we have in our stockpile and how quickly and easily can they be manufactured? The answers are probably "not enough" and "not fast enough"!

GPS has been so successful that other countries are now building their own satellite systems to provide similar services. Why? If the American system is in place and available free of charge, why would anyone invest the money to build a network of satellites providing a duplicative capability? To answer the question, all one has to do is ask another "Would you want all your systems that depend upon GPS to be shut down if the United States decides to deny access to it?". I suspect not, hence their interest in building a redundant backup system.

The Russian system, known as the Russian Global Navigation Satellite System, or GLONASS, was made available for civilian users in 2007. The Europeans, Chinese, and Indians are each either building or planning to build their own satellite navigation system using independent satellites.

I'm not claiming to be the first person (not even close) to contemplate the military implications of losing our GPS and other global positioning satellite systems. Just as potential adversaries realized the incredible advantage provided to the United States by space-based systems during the Gulf War, so did our own leaders – and they also began to consider what might happen if that capability were denied to us.

In a February 2010 speech, Air Force Chief of Staff General Norton Schwartz made the following statement:

> "Our dependency on the Global Positioning System has also created certain vulnerabilities that our adversaries can exploit through jamming and other tactical denial techniques. While we remain unequivocally committed to proper stewardship and use of the world's unparalleled standard in precision navigation and timing, as well as advancing enhanced capabilities with new GPS Block IIF satellites and next-generation GPS III concepts, *we also recognize the need to be able to continue to operate effectively, through improvement to GPS and other methods, in a denied or degraded localized environment.*"
>
> Jules McNeff, "Military PNT – The Way Ahead: Changing the Game Changer", *Inside GNSS*, November/December 2010, *www.insidegnss.com/node/2347*, italics added by the author

Reference

[1]   US Congress, National Defense Authorization Act for Fiscal Year 1994, P.L. 3-160, Division A, Subtitle D, Section 152(b). See also Senate Armed Services Committee Report 103-112.

5 Economic Fallout

The idea for this book originated after I stopped at a local supermarket on the way home from work to pick up something for dinner. The stop was intended to be quick. I needed only a few items and I was anxious to get home and be with my family.

As I walked into the store, I recall noticing that the checkout lines were longer than usual. This didn't really surprise me, as many busy office workers stop at the market on their way home to do just what I was doing. I didn't really think much more about it until I was ready to check out.

The lines had gotten longer, not shorter, and they didn't seem to be moving. After a few minutes of standing and not moving an inch, I began to get curious as to why. I looked at the other lanes to see whether I had just picked the wrong one (in which someone was waiting on a price check or something else equally time-consuming) and weighed the option of jumping to another checkout line. None of the other lines was moving either. Frustration mounted.

The store manager was moving from line to line and he finally made it to ours. He looked a bit frazzled.

"We're having a problem processing credit cards and checks. Please be patient while we try to get the problem fixed," he said.

After a few minutes that seemed like an hour, the manager came by again. This time, he was holding a cash drawer in his hands.

"Our satellite connection is down and we can't process any credit cards or checks. We don't know when it will come back online so we are opening another line – cash only. I'll have it up and running in just a few minutes. If you will be paying in cash, please come to register one." He then walked over to an unused lane and prepared to open it.

Surprisingly, I was one of the few people who moved to the new "cash only" line. I had only a few items and the money in my wallet was more than sufficient to pay for them.

I'm very curious about all things space and space-related (one of the hazards of working for NASA). Since there was no one else in the line with me, I took the opportunity to ask the manager about the problem.

"All of our credit card transactions and check verifications are done by satellite. If you look outside, you'll see the antenna on the roof," he said in response to my questions. "The connection goes down sometimes, but it usually comes back up faster than this. I'm really sorry …..''

As I walked out of the store I looked at the roof and, sure enough, there was a satellite dish there, pointed toward the southern sky. I then looked at the gas station next door – another satellite antenna. I looked at the gas station across the street – there was yet another one. On the way home, I drove by some other stores and noticed that almost all of them had satellite dishes. This was beginning to get interesting ...

Very Small Aperture Terminals

The dish antenna at my supermarket was most likely a VSAT, or "Very Small Aperture Terminal". Many large and small retail chains use VSATs to conduct financial transactions (like my recent purchase) and to update their inventory levels. VSATs offer a fast and more direct method to conduct a sale (gaining credit card approval, updating inventory, etc.) and have now become an integral part of the American retail industry. A typical VSAT is pictured in Figure 5.1.

As I noticed driving home that day, most gas stations are also VSAT users. A list of other major VSAT users can be found in Table 5.1.

Table 5.1. VSAT systems are being used all over the world by private companies, governments, and financial institutions (data courtesy of Comsys).

VSAT user	Country	Approximate number of sites
Automatic Data Processing, Inc.	USA	2,172
Albertsons	USA	1,586
Bank of China	China	340
Best Western Hotels Canada	Canada	168
Dunlop	Nigeria	30
ExxonMobil	Malaysia	600
GESAC	Brazil	10,520
National Stock Exchange	India	5,950
State Farm Insurance	USA	33
Yum! Brands	USA	3,623

Among additional VSAT users in the United States are Wal-Mart (2009 annual revenue ∼ $400 billion), Walgreens Pharmacy ($63 billion), CVS Pharmacy ($37 billion), and even fast-food restaurants like Taco Bell ($1.9 billion). Today, there are nearly 750,000 VSAT terminals in use worldwide [1].

It turns out that the disruption at our local supermarket was local – limited to just that one store. But what would have happened had it been global? What if some or all of these major retailers were to lose their satellite connections at the same time?

9/11 was a shock to the US economy. For about a week, many Americans stopped shopping, many didn't go to work, and those who did were not very productive as discussion of the tragedy took priority over what was going on in the workplace. Slowly, as people realized there were not going to be any more immediate attacks, daily living – and shopping – resumed. The economy may have sputtered, but it didn't crash. We moved on.

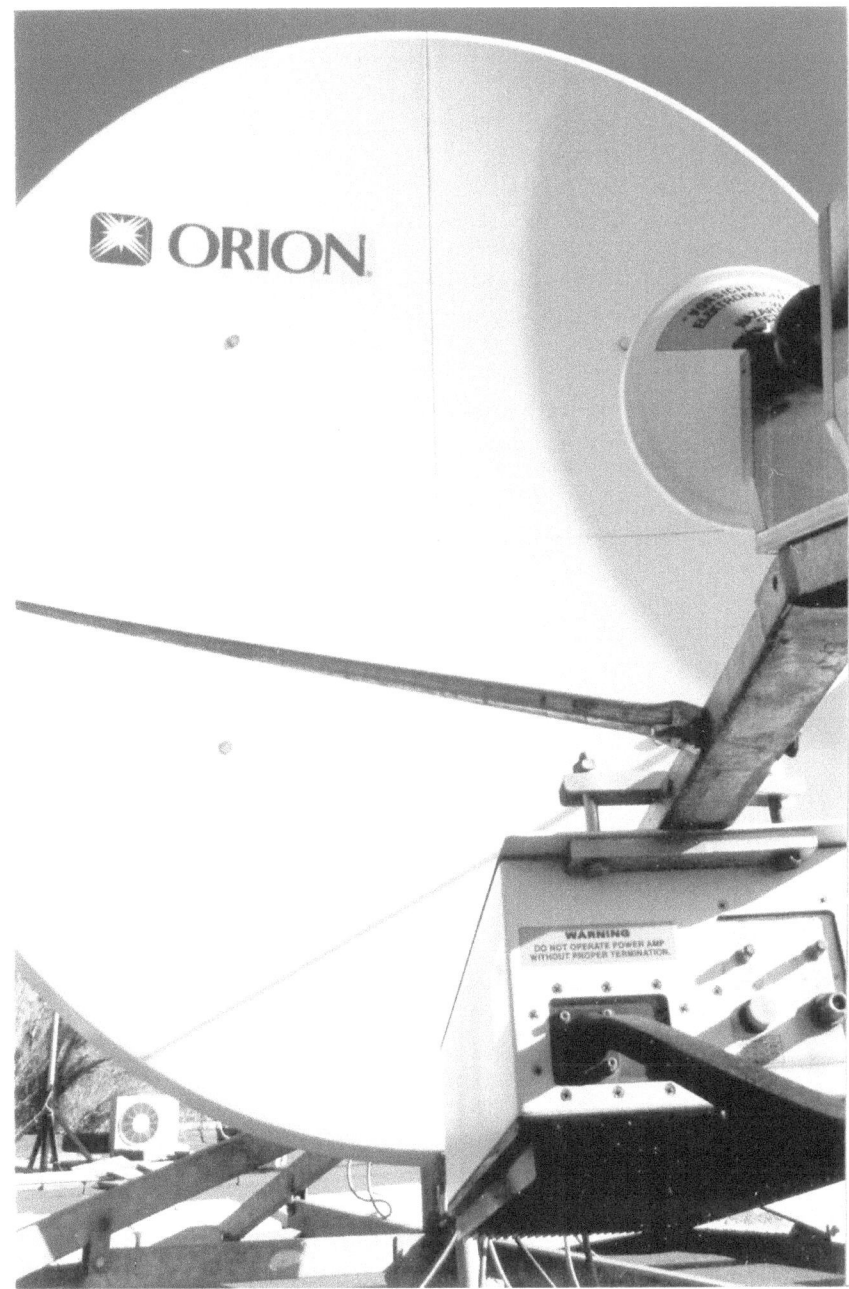

Figure 5.1 A typical VSAT terminal. This 2.5-m parabolic dish antenna is used for high-speed satellite internet communications. (Image courtesy of Adamantios)

The loss of satellite communications in the retail segment of our economy would not cause a singular, catastrophic economic crisis like 9/11. Rather, since it would likely play out over weeks and months, the resulting negative economic impact would be gradual. At first, there would be a widespread loss of sales (revenue) as consumers found themselves unable to use their credit or debit cards. Many would switch over to cash fairly seamlessly. For others, probably many others, abandoning their credit cards and using cash would cause a major cash flow problem. For them, purchases would be postponed until payday and many would simply decide to not buy at all.

When significant numbers of consumers stop making purchases, the result is a recession.

Recall that VSATs are also used by retailers to manage their inventories. When you purchase something at a major retail chain, they use the bar code on the merchandise to record the sale price – what you pay – and to keep track of what they've sold – their inventory. A central computer usually then gathers this information from multiple stores and uses it to reorder the products that are selling well. The end result is that the retail store rarely, if ever, has empty shelves. When was the last time you walked into a major retail store and saw empty shelves labeled with a sign saying "Out of stock"? Almost never. An empty shelf is lost revenue. And an empty shelf is avoided with a rapid, mostly automatic merchandise reorder system – like a network of computers connected to each other with VSAT terminals.

Now, back to your local retail store. In a worst-case scenario, customers would have been buying products with cash and there would not have been any inventory data flowing out in the weeks or months since the satellites went away. Many shelves would be empty – but only of those products you need the most. After all, that's why people have been buying them!

Would the stores recover? Yes, retail stores did fine in the age before satellites and they would be able to do so again. But it would take time and investment to put an alternative in place, probably one using an internet-based solution.

With 70% of the American economy depending upon consumer spending, a global disruption in buying would likely cause a recession. At worst, it could cause a depression.

Would other segments of the economy be negatively affected?

Air Travel

Delta and American Airlines, the two largest American airlines, have a combined revenue of about $47 billion. Airlines have operated successfully since long before satellites were put into space, carrying millions of passengers all around the world in safety and at reasonable cost. Would they be affected by a loss of space satellites?

Yes. In just a few years, global commercial aviation will be totally dependent upon Earth-orbiting satellites. Why?

The Next Generation Air Transportation System, or NextGen

With the growth of air travel, the current air traffic control system of land-based beacons, radar, and voice communications is strained to the maximum and would be very difficult, and expensive, to upgrade. The solution is NextGen. NextGen is based on Global Positioning System (GPS) satellites in low-Earth orbit.

If each aircraft can precisely know its location and if every air traffic controller can use those data without having to rely on radar interpretation and human error in voice communications, then the result would be a more accurate and safer air traffic control system. GPS can provide this information and most in the airline industry are looking forward to its implementation. GPS navigation will make their jobs easier and allow for continued growth in the industry.

When NextGen is in use, aircraft will depend upon it to navigate. It is inevitable that the existing network of beacons and radar installations will fall into disuse and eventually be dismantled. After all, who wants to pay for maintaining a redundant and costly backup system? Not cost-conscious airlines and certainly not cash-strapped governments. In the near future, the commercial aviation industry will come to rely on GPS.

And what if GPS then disappears? For safety, the airlines would have no choice other than to ground their fleets. To understand why, one has to look no further than the Hartsfield-Jackson Atlanta International Airport in Georgia. According to Airports Council International, Hartsfield-Jackson has been the busiest airport in the United States every year since 2000. In 2009, the airport handled over 970,000 flights. That's between one and two flights every minute. The current air traffic control system is straining to handle the load and the planned NextGen will make it much easier – and safer – to do so. But, after NextGen becomes the only method for managing nearly a million flights per year (at a single airport), the system will break without it. To the best of my knowledge, there are no plans to maintain a non-NextGen backup system to cope with this potential scenario.

According to the Federal Aviation Administration (FAA)'s NextGen website, civil aviation comprises approximately 5.2% of the US Gross Domestic Product (GDP), contributing $1.3 trillion dollars to US economic activity and directly employing 10.2 million people.

Rail

Like the airlines, North American railroads will soon implement a GPS navigation and tracking system called Positive Train Control (PTC) [2]. Under PTC, each train will determine its exact location, communicate that to a central control center, determine, in a mostly automated system, whether its movement is safe with regard to the locations of other trains, and then either allow or disallow travel.

PTC was proposed and signed into law in part as a response to a collision between a passenger train and a freight train in 2008. The new system would prevent such collisions by requiring that every train's position to be known, allowing real-time collision avoidance. Like airlines, if the GPS system were to go down once PTC is implemented, rail traffic would stop.

According to the American Association of Railroads, railroads sustain 1.2 million jobs. Each year, railroads carry over 1.5 million carloads of agricultural products and another 1.5 million carloads of processed food. They also carry millions of carloads of cars, paper, lumber, and the raw materials needed by our factories to keep us supplied in our homes and businesses. Most significantly, railroads account for more than 70% of coal deliveries to the nation's electrical power plants [3]. If the flow of coal stops, then so does the delivery of electricity for about half of the population.

Manufacturing

Once upon a time (25 years ago), manufacturing was not nearly as globalized and competitive as it is now. Businesses that made things, widgets, would order the parts required to make their product and store them in large warehouses until they were needed. When supplies started to get low, the business would reorder and continue to use their stockpiled parts inventory until the new parts arrived. If something happened to interrupt the supply chain, there was typically time to adjust and find alternative suppliers. Sometimes, these businesses would stockpile months of parts at a time.

Along came several major changes to the business world that dramatically affected this approach to manufacturing: intense competition from companies in multiple countries making comparable products at a lower cost; computerization, which allowed tracking of inventories accurately and quickly (not only the computers, but also bar codes, radio frequency identification, etc.); precise navigation (GPS); and rapid communication (cell phones, satellite phones, and the internet).

To effectively compete, most large manufacturers adopted *Just In Time* inventory systems [4]. The premise of *Just In Time* is simple: keeping a large inventory of supplies is expensive and it doesn't allow companies to quickly respond to market changes. Instead, it is much less expensive to keep minimum inventory, with component parts being constantly in transit to the manufacturing facility, arriving just as they are needed on the assembly line. The primary focus of *Just In Time* inventory is having "the right material, at the right time, at the right location, and in the exact amount needed" [5].

Why is this generally better for a business's bottom line? If the demand for a particular company's product dips, keeping several months of spare parts in a nearby warehouse might turn into an expensive liability should demand not quickly pick up again. If a company practices *Just In Time* inventory, then they can simply not buy the next shipment of parts until it is needed.

Unfortunately, this makes businesses entirely dependent upon a complex supply chain and the logistics thereof. If anything disrupts that chain, then the business will not have the supplies it needs to make its products, shutting them down.

Where do satellites figure into this supply chain? Everywhere.

In our global economy, a manufacturer in Chicago may assemble the finished product that will ultimately be sold to consumers, but where are the component parts made? The simple answer is – somewhere else. The parts might be made elsewhere in the United States, or in Canada, Europe, Japan, or China. Without GPS, the airplanes that ferry parts across the globe will be grounded and the massive sea-going cargo ships will have to find alternative ways to navigate – returning to the much less efficient techniques used decades ago.

The VSAT systems used by retailers to manage store inventories are also used by manufacturers to manage their inventory of parts and supplies, keeping factories that depend upon one another using *Just In Time* practices with the supplies they need, when they need them. If the VSAT systems go down, then so does the inventory management system that keeps manufacturing viable under *Just In Time*.

It is also worth pointing out that manufacturing is not the only industry that uses *Just In Time* inventory management. How do you think your local gas station maintains a constant supply of fuel for your car? Tanker trucks are constantly in transit from the refinery and storage facilities to gasoline stations across the country. The very same technologies – GPS, satellite and cell phones, and VSAT terminals – are essential in maintaining this supply chain. With a major disruption, businesses and individuals might, quite literally, run out of gas.

Our modern industrial base is more efficient and productive than it has ever been thanks to modern computer, satellite, and communications technologies. It is also more vulnerable to disruption. Many of the tools that enable *Just In Time* to function are dependent upon our satellite infrastructure being reliable and available. Recovering from the loss of this infrastructure would be expensive, painful, and slow.

The Fishing Industry

Fishing is one of the oldest industries in the world and it has gone high-tech. Ships no longer leave port for parts unknown, hoping to catch fish and bring them home to the local fish market for sale. Fleets of ships now go to sea, carefully navigating to assure that they remain in areas approved for fishing by the world's governments, catching only the numbers they are approved to catch, and following the prevailing market conditions that will determine when and where their catch is delivered for sale and processing on the way to your local supermarket shelves.

For example, European fishing ships are required to register and report their voyages electronically, apprising government regulators of where they fish, and what and how much they catch. The two most common approaches used by fisherman to meet these regulations are Iridium-based satellite phones and VSAT terminals aboard ship. In addition to complying with regulators, fishermen now use these systems to get weather information and fuel costs and to find out at what port they can get the best prices for their catch. And, yes, most fishing ships now use GPS for navigation as well.

Satellite remote sensing, using scientific data collected from space satellites looking at the world's oceans, has increased fishing productivity and is helping to prevent over-fishing in selected areas. Near-real-time pictures from space, combined with GPS, can help fishing vessels locate schools of fish [6]. As a specific example, to determine the best place to go for a day's catch, fishing fleets use satellite-determined ocean temperatures, which have been correlated over time with the location of schools of tuna.

With modern technology at their fingertips, fishermen can precisely determine where the fish are located and navigate to them so as to catch as many fish as possible. Without these data, the enterprise becomes much less productive, resulting in far fewer fish being caught and a dramatic reduction in commercially available fish and fish products. This multi-billion-dollar industry has been transformed by satellite technology and our diets, as well as the industry itself, would suffer greatly if the fishing industry's productivity were to be slashed by the loss of these vital tools.

The impact on the global economy from a reduction in fishing production would be widespread. Not only would there be fewer fish in the grocery store, but restaurants would suffer immensely. Just one species, the pollock, is widely used by fast-food chains like McDonalds to make their fish sandwiches and by other companies in their frozen fish meals. Consumers are buying fish oil supplements at an accelerating rate and all of the businesses associated with this depend upon more fish being caught each year. Without satellites, there is simply no way we could maintain current production levels, let alone increase them.

Television

And now to the more mundane: television. Don't mess with people's televisions. It was on television that we watched the Twin Towers fall. It was there that we learned that New Orleans flooded and of the London subway bombings. We turn to our televisions to learn the latest news during a crisis. And, in this particular crisis, television would be of no use whatsoever.

As explained in Chapter 7, cable television, which serves about 58% of the US population, would become nothing but static if our satellites were to be destroyed. Your local cable service provider gets its 200 channels of news, sports, and entertainment from satellites that relay signals to them from anywhere in the world. If all of our satellites are destroyed, then this industry will virtually cease to exist.

Local broadcast television stations will not fare much better. They, too, get nearly all of their programming from satellites. Network news? Gone. Late-night

Figure 5.2 An artist concept of the Space Systems/Loral Amazonas 3 communications satellite. (Image courtesy of Space Systems/Loral)

talk shows? Gone. Your favorite soap opera or National Football League (NFL) game? Gone. If the network feeds go down, then only local programming would remain – for a very long time. Shown in Figure 5.2 is an artist concept of the Space Systems/Loral Amazonas 3 geostationary communications satellite. Amazonas 3 has a wingspan comparable to a jet airliner, weighs 5 tons, and has over 50 transponders – typical of the satellites in geostationary Earth orbit (GEO) proving television signals to virtually any country in the world.

So, other than forcing people to be more sociable and perhaps healthier because they are no longer able to be sedentary while watching their favorite reality show, how, one might ask, would the loss of television be a bad thing? After all, haven't educators been decrying the boob tube for more than a generation?

Television advertising is a $40–50 billion per year industry. Cable television revenues (from subscribers' fees and other non-advertising sources) add up to about $20 billion per year. This $60–70 billion per year in large part become employee salaries. If the programming cannot reach the customer, then the ad revenue will fall. If advertisers don't advertise, then sales will fall. If the actors and actresses in these hundreds of television shows have no work, then neither do their script writers, producers, set designers, and other production team members. The result is an economic domino effect from industry to industry.

Satellite Radio

A relatively new entrant in the satellite communications industry is satellite radio. Until recently, there were two competing companies, Sirius and XM, who have now merged. Satellite radio offers a suite of special-interest radio channels available only to paid subscribers via specialized radios. The radio signal is beamed to them from satellites in GEO. Among the benefits of the service is a continuity of programming no matter where you are. If your interest is classical music, then you may set your receiver to a favorite classical channel and listen to it, nonstop, as you drive from New York to Chicago and then on to Los Angeles.

Needless to say, this is one industry that is totally dependent upon satellites.

The Cell Phone Industry

Cell phones are not satellite phones. Cell phones, in most cases, do not need to relay information to and from a satellite when you are trying to call your mother on Sunday evening. A cell phone is nothing more than a sophisticated radio transmitter and receiver. And, as with most radios, the distance across which it can transmit is limited by the power and capability of the individual phone. It is for this reason that cities, major highways, and, increasingly, rural areas are divided into regions, or "cells", that allow for individual cell phone frequencies to be received and retransmitted to other cells until the signal finally reaches its intended recipient. The signal goes from tower to tower on the way to its final destination, in most cases never leaving the ground network. (There are exceptions for international and other very-long-distance calls. Some of these calls do, indeed, travel to and from space satellites.)

Unfortunately, the cell phone system is not immune from a loss of space satellites. Keeping up with, not mentioning routing, the billions of cell phone calls and text messages originating in the United States each year is a daunting task. To make this tracking and routing possible requires precise timing. The time stamp for each packet of cell phone data must be known and agreed to by all the towers and transmission stations between you and the recipient of your call or text. It is for this reason that cell phone companies use GPS receivers at their base stations and towers. These cell phone base stations are synchronized by a GPS signal. GPS provides the required precision timing and, thanks to satellite technology, a synchronized timing signal for base stations around the world.

If the timing signal is interrupted, the system can no longer route calls. If the GPS system goes down, then that cell phone in your pocket may become nothing more than an expensive paperweight.

According to the wireless industry association, CTIA (Cellular Telephone Industries Association), there are over 300 million cellular subscribers in the United States, roughly a third of whom have no landline service. This is over 80% of the population. CTIA estimates that cellular revenues in the United States alone were over $160 billion in 2011. And that is just the direct impact of cell phones on the economy. How many businesses depend on cell phones and, increasingly, on smart phones to conduct their day-to-day business? The number is impossible to know with any accuracy, but it most likely dwarfs the actual revenues of the cell phone industry itself. According to the United Nations, there are now over 4.6 billion cell phone users worldwide.

If the cell phones stop working, then so will businesses and many aspects of government.

I am really not trying to be alarmist. The truth speaks for itself and the conclusion is scary. Our economy – our way of life – is now more dependent on highly specialized technologies than ever before. We are aware of our dependence on oil and other fossil fuels. The "energy crisis" and the threat to our economy from a potential loss of oil is constantly in the news and you would have had to have been living under a rock for the last 40 years to not be aware of it. The threat to our satellites, which is just as real and has been looming larger over the same time period, has been largely ignored.

Our economy has grown dependent on and accustomed to the benefits provided by satellites. They are taken for granted in much the same way as we take for granted the anticipated rising of the Sun every morning. Unfortunately, there is a very real possibility that they may not always be there for us. And, if they were to go away, the economic consequences could be significant.

References

[1] www.comsys.co.uk.
[2] US Department of Transportation, Federal Railroad Administration. *Five-Year Strategic Plan for Railroad Research, Development, and Demonstrations.* Washington, DC (2002).
[3] Association of American Railroads. *The Economic Impact of America's Freight Railroads.* Washington, DC (June 2012).
[4] Hutchins, D. *Just In Time.* Gower Technical Press, Ltd, Aldershot, Hants, England (1999).
[5] Production & Operation Management. StudyMode.com.05 (2011).
[6] Kapetsky, J.; Aguilar-Manjarrez, J. Geographic Information Systems, Remote Sensing and Mapping for the Development and Management of Marine Aquaculture. Food and Agriculture Organization of the United Nations, Rome, Italy (2007).

6 The Global Positioning System and the Average Person

If you need help navigating in a strange city, you are likely to use your Global Positioning System (GPS) receiver. For under $200, you may purchase a fully functional GPS receiver that will help you get safely from Point A to Point B. They are now a standard option in many rental cars. Have you bought a house or some land recently? Chances are that a survey was performed using GPS's ability to provide highly accurate position knowledge as a part of the process.

Scientists use GPS data in their study of earthquakes and of the Earth's atmosphere. It is difficult to find a realm of modern life that is untouched by GPS. How would we adapt if GPS were to disappear?

There is some debate as to when the first commercial vehicle GPS navigation system was made available to the general public for purchase. However, virtually all of the claims put the date in or near the mid-1990s. By 1995, automotive companies like Oldsmobile were making on-board satellite navigation by GPS an optional feature on their higher-end cars. In the year 2000, the US government made a more accurate GPS signal available for general use and the number of cars with satellite navigation increased dramatically. Today, if your car doesn't have such a unit built in by its manufacturer, then you are likely to have a stand-alone unit sitting on your dashboard that you can purchase virtually anywhere or online. With accuracies of a few meters, and with continents' worth of updateable maps installed and the ability to upgrade the maps as roads change, these devices are indispensable for many of today's drivers – commercial and personal. I use mine whenever I travel to another town and often to find locations with which I am generally unfamiliar within my own community.

When you rent a car in a strange city, the rental car company will offer you the option of adding a GPS unit with the car. (Personally, this is an option I usually decline. I prefer to use the GPS features built into my smart phone.)

I have two children, both of whom are now teenagers and new drivers. They are not proficient at reading maps. But they can certainly enter the address of where they want to go and follow the voice commands from "the Garmin Lady" that inhabits our car's dashboard. (My daughter gave her an English accent – more *Harry Potter*-ish!) I am remiss in not making them learn to read a map; I suspect they would have difficulty recalling that the top of a page is usually north. Alas.

And then there are *Google Maps*, *Mapquest*, *Rand McNally*, and others who develop and post precise and accurate maps online for helping people navigate –

and, of course, these maps and their users rely on GPS data for accuracy and to get them from one place to another.

And everyone seems to be using GPS. For my job, I sometimes travel internationally. On a recent trip to South Korea, I was astounded by the penetration of GPS in virtually every car, taxi, and bus we encountered. My host's GPS unit had a split screen that enabled him to watch local television broadcasts while navigating – an obvious safety issue about which I did not comment.

When I find myself in large cities, I use the GPS in my smart phone to figure out which way I need to walk in order to get where I want to go. All smart phones now have this feature and I suspect many people are like me, who cannot now imagine ever having to navigate without it again.

Even phones which don't have GPS mapping capability have GPS, at least in the United States. In 2005, a new law took effect that requires all new cell phones to have special software or chips installed and functioning so that emergency services can locate a caller without having to ask for directions. The law is called "Enhanced 9-11" or "E-911" and is designed to make calling for emergency help faster and easier in order to save lives. While not all E-911 calls are geolocated using GPS, the trend is definitely towards this approach.

In critical emergency situations, these GPS-enabled phones can be tracked without the carrier even placing a call. This feature has been used to track missing persons, lost or abducted children, and disaster victims. Emergency responders now use it routinely.

GPS is now also being used by law enforcement to track the whereabouts of suspected criminals, although the practice is not without limitations – at least in the United States. The US Supreme Court recently ruled that using GPS tracking devices by law enforcement would require a court order, thus overturning a man's criminal conviction on drug-related charges.

GPS receivers and smart phones can do far more than tell you where you are. Are you interested in finding a good place to eat Thai food for dinner? If so, then check the list of restaurants downloaded into your receiver or use one of the many smart phone apps that will find your current location and then recommend places to eat in close proximity based on other users' reviews. These apps may be free or nearly free to download, but don't believe for a minute that the creators are not making money from them. Advertisers and other users are paying to have their business listed. Apps are now big business.

If you are a boater, then you realize the importance of reading navigation maps and understanding where you are relative to underwater threats like large debris or even just a sandbar. It is vitally important for you to know where you are relative to any of these potential hazards. Companies now make GPS chart plotters that will show your location on a maritime map, noting the proximity to known hazards and allowing you to compare your travel plans with required navigationally safe routes.

As of this writing, the investigation into the causes of the sinking of the Italian cruise ship *Costa Concordia*, which killed at least 30 people, is not complete

Figure 6.1 The cruise ship *Costa Concordia* sank in well-known waters, killing at least 30 people, despite the availability of GPS tracking. (Image courtesy of Roberto Vongher)

(Figure 6.1). From news reports, it appears that that the accident need not have happened. The waters in which the ship was sailing were well mapped and the crew should have known their exact location using GPS receivers on board the ship. This tragic accident happened despite the functionality of GPS. How many commercial crewmembers know how to navigate without it?

There are now sports and games that require GPS. I was hiking in our local state park a couple of years ago and came across a strange logo on a rock. At about that time, another hiker came through, saw my puzzlement, and then proceeded to tell me all about how active our local geocaching group is. The symbol on the rock was a stylized X within a circle, making it appear to be a circled letter "G". In the game, someone leaves a waterproof container at a geographically known spot. The location is, of course, determined by GPS. Within the container is usually a logbook in which all those who find it are asked to enter their name or codename. There may also be other items included in the box, but that isn't required. The geographic coordinates of the box are made public and any who wish to play then go out and use their GPS to find the boxes, signing the logbook when they do.

Cameras that take advantage of GPS are now automatically noting the location where photos are taken (geotagging). Using the data, photographs can now be easily organized digitally, allowing biologists to study specific plants or animals in a geographic region with the click of a mouse. With the date and

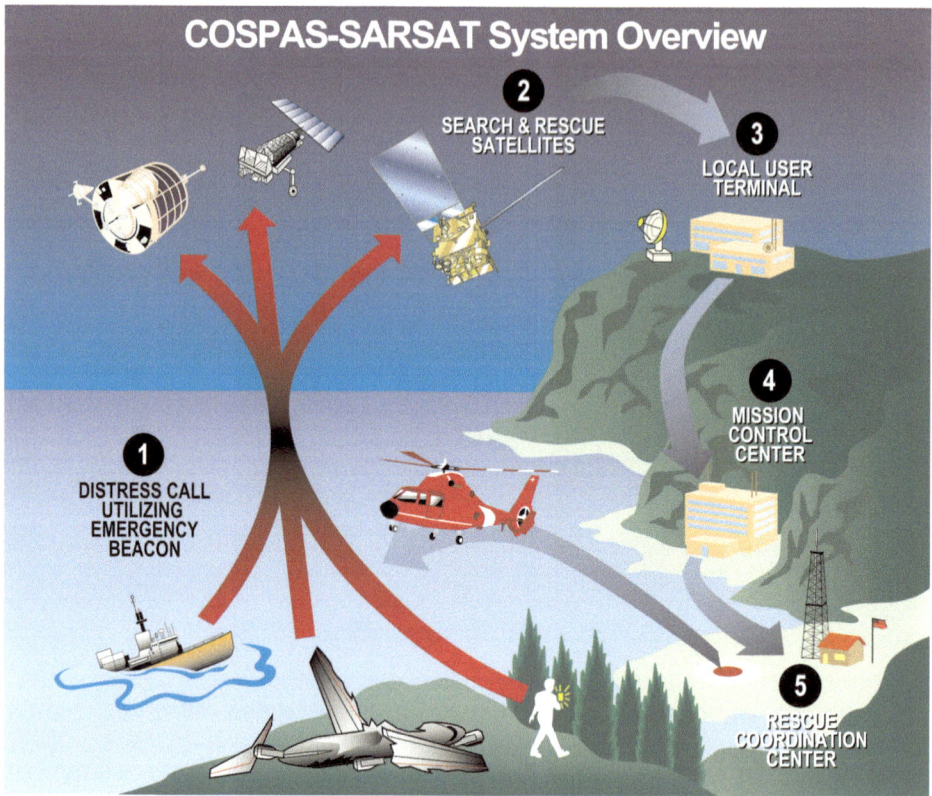

Figure 6.2 The COSPAS–SARSAT system is an international program of search and rescue that depends upon satellites for saving lives. (Image courtesy of NOAA)

location encoded in the digital photograph, pictures can now be easily sorted in whatever way best answers the scientists' desires.

The US NOAA, in collaboration with Canada, France, and Russia, operate the Cosmicheskaya Systyema Poiska Sudov (COSPAS)–Search And Rescue Satellite Aided Tracking (SARSAT) system to detect and locate distress signals from emergency beacons virtually anywhere in the world, providing the data to emergency rescue teams from many nations who can then provide assistance to those in distress. The SARSAT system uses GPS and other satellites in low-Earth orbit (LEO) and geostationary Earth orbit (GEO) to detect an emergency beacon, relaying the signal to the nearest node within a network of ground stations around the world when it is activated. The ground station provides the location information from the beacon to the rescue team (Figure 6.2).

The SARSAT has an interesting history that began in 1972 when an airplane carrying two US congressmen disappeared over a remote region in Alaska. Despite a massive search-and-rescue operation, the congressman and their plane were never found. As a response to this incident, the US Congress mandated that all aircraft in the United States be outfitted with an Emergency Locator Terminal (ELT), which would turn on in the event of a crash and act as a broadcast locator beacon. Over time, the technology improved, became digital, and then satellite-based – providing global coverage. According to NOAA, there are now more than 350,000 beacons registered and in operation. The program has an astonishing success rate, with over 6,110 people rescued in the United States since 1982 [1].

The list goes on. GPS is almost as ubiquitous as electricity and its loss would be immediately felt by just about everyone reading this book.

Reference

[1]   *www.sarsat.noaa.gov/cospas_sarsat.html.*

7 Spy Satellites and Military Communications

Control the high ground and you have a good chance of controlling the outcome of a battle. This fact has been known to soldiers since the earliest days when the high ground was the hilltop upon which communities would build their homes – making them defensible during an attack and affording them with a high probability of seeing the advance of a possible adversary long before they were close enough to be a real threat.

Control the highest point on a battlefield and your arrows will fly farther and your cannonballs can be lofted to engage the enemy before theirs can reach you. You can also see the enemy at a distance, which often gave those on the high ground time to prepare for battle, making them less likely to be caught by surprise.

Today, the United States and its allies essentially control the high ground of space. We base our military doctrine and war plans assuming that our satellite communications, cameras, and GPS signals will be available and working correctly in all of our future conflicts. This is a dangerous assumption that may lead to what some have called a "Pearl Harbor in space". To understand why we are in this position, we need to discuss the history of airborne and space operations.

Before the airplane, armies experimented with balloons to loft soldiers higher into the sky than the local mountains so as to enable them to see what their enemies were up to. Some might have even flown directly over enemy encampments, taking note of the activities below so they could report back what they'd learned. This assumes, of course, that their balloons could fly high enough to be out of range of whatever weapons those whom they were observing might have handy.

With the development of airplanes came airborne reconnaissance. Pilots would fly over their adversaries, both real and potential, taking pictures and reporting their findings back to their superiors by radio. As the airplane technology improved, so did the ability to spy on targets farther and farther away from home. The Second World War gave birth to the modern notion of aircraft reconnaissance and the Cold War gave us a chance to perfect it. Unfortunately, long-range weapons technology advanced as rapidly as the aircraft technology used for spying. While many reconnaissance aircraft were lost during the Second World War, it was not until the fragile peace of the Cold War that this reality hit home.

During the mid to late 1950s, the United States developed and flew a high-altitude aircraft known as the U-2 (Figure 7.1), which routinely over flew the

Figure 7.1 A U-2 spy-plane similar to the one flown by Francis Gary Powers in 1960. (Image courtesy of the US Air Force)

Soviet Union at altitudes above 21 km – out of reach of their air defenses at that time. Needless to say, the Soviets were not happy about the situation and they worked feverishly to develop a capability to shoot these aircraft down. They succeeded in 1960 when they shot down an aircraft piloted by Francis Gary Powers, sparking an international incident.

Aircraft reconnaissance technology continued to evolve out of necessity. In a world in which the destructive capability of modern armies literally threatened the existence of entire countries and perhaps life on the planet, knowing what your potential adversaries were up to could make the difference between survival and destruction. In addition, as the world's superpowers were negotiating arms-control treaties to limit their respective numbers of nuclear weapons and their ability to test them, being able to verify that the other side wasn't cheating was an important thing to do. Without the ability to see what each side was doing, situations might arise leading to misunderstanding or perhaps accidental war.

The spy-plane game continued to evolve, with each side developing planes that could fly faster and higher than those before them. From this arms race came the SR-71, capable of nearly reaching space by flying at altitudes above 25 km at speeds in excess of 3,200 km/hr [1].

The game changed when the Soviet Union launched Sputnik, the world's first satellite, in 1957. Flying in an elliptical Earth orbit at altitudes of between 938 and 340 km, the Sputnik demonstrated a capability to fly over virtually any part of the planet with impunity. A new high ground was seized and space-based reconnaissance was born.

The United States responded in 1958 with the launch of Explorer 1 and soon both countries were filling nearby space with satellites [2].

Based on what has been declassified, which may not be everything in this time period, the United States' first spy satellites were part of the Corona Project, which operated from 1959 until the early 1970s. The Soviet Union, of course, had its own spy satellite program, called Zenit. Both Corona and Zenit satellites flew high-resolution cameras using photographic film that had to be recovered and developed for study. Hundreds of these satellites were launched into space, each peering down at military installations all around the world.

Technology continued to improve, eliminating the need for spy satellites to carry photographic film (with the advent of digital imaging). Today's spy satellites can take pictures of objects on the ground with almost unbelievable clarity (Figure 7.2), making it virtually impossible to conceal any sort of military operation, facility, or encampment.

Let's talk about space satellites and observations in the context of a nuclear-armed world. During the height of the Cold War, the United States had thousands of strategic nuclear weapons. (In this sense, "strategic" can be distinguished from "tactical" by their intended purpose and total destructive power – tactical nuclear weapons are to be used on a battlefield in a direct

Figure 7.2 This 0.5-m-resolution satellite image shows Tokyo, Japan's central business district, and Shibuya Station, one of Tokyo's busiest railway stations. The image was collected by the GeoEye-1 satellite on 15 May 2010, while flying 680 km above the Earth. (Satellite image by GeoEye)

military exchange, whereas strategic nuclear weapons would be typically used to inflict massive damage on a country's military infrastructure and would be launched by a missile or aircraft.) The Soviet Union held a similar number. The Cold War avoided becoming a "hot war", meaning that actual shooting would begin, in large part due to the notion of deterrence. Both sides had so many nuclear weapons that it would be unthinkable for either side to start a war because both sides would inevitably lose. A nuclear exchange would be devastating to both countries and just about everyone else in the world.

Thanks to spy satellites, both the United States and the Soviet Union knew where each other's nuclear weapons were located and, since a missile can cross the globe in less than 90 min, both were vulnerable to having their missiles destroyed before they could be launched in a retaliatory strike. If one side felt they could launch enough nuclear-tipped missiles to totally destroy the opposition's ability to respond in kind, then they could, in theory, "win" a hot war by striking first. Military planners considered this scary scenario a very real possibility for several decades.

To mitigate this risk, new generations of spy satellites were developed to watch for the telltale signs of rocket launches that could herald the beginning of a nuclear first strike. If the satellites detected multiple rocket launches that were thought to be nuclear-armed missiles, a country had basically two choices:

1. Wait until the nuclear weapons on the missiles struck their targets, to make sure they were, in fact, nuclear missiles and not something else. This also allowed for an assessment of where the missiles were headed. It was possible, though not likely, that the target might be someone else. The problem with this approach was that if it was, in fact, a nuclear attack, you might lose so many of your nuclear weapons in the attack that you would be unable to effectively respond. In this unlikely scenario, the "other side" would win.
2. Launch immediately, before the missiles struck, to make certain that the opposition would suffer extreme damage as a result of their sneak attack. The obvious problem with this scenario is that the attack detected by the satellites might not really be an attack on your country and, by launching so soon, you might be the one to start a nuclear war. With missiles crossing the globe so quickly, a political leader would not have much time to make a launch decision.

To watch for potential nuclear attack, the United States developed and launched the Defense Support Program (DSP) satellites (Figure 7.3). The DSP satellites were deployed in geosynchronous orbit above the equator. They used infrared sensors to detect the characteristic heat of a rocket's exhaust, providing the early-warning capability required. The first DSP satellite was launched in 1970 and the final one flew in 2007. A new satellite called the Space-Based Infrared System (SBIRS) is replacing the DSP satellites [3].

Though the Cold War is thankfully over, many countries still use satellite-based missile launch decisions based on near-real-time information from

Figure 7.3 The Defense Support Program (DSP) satellites provided early warning of rocket launches anywhere in the world. (Image courtesy of the US Air Force)

satellites. The United States and Russia still maintain a sizable nuclear deterrent; although thankfully much smaller than that held in the Cold War, it is still enough to destroy each country many times over. The militaries are still watching each other, making sure that the political détente and overt cooperation between them are not a façade, waiting to be broken in some sort of nuclear sneak attack.

If our spy satellites were to stop working, then our ability to know what was happening on the other side of the planet would go away. Military planners in both countries would suddenly have to ask the question, "What if the other side uses this time of blindness as an opportunity for a sneak attack?" Would one side or the other ponder such an attack? Perhaps equally importantly, would the other side think it was a real possibility and launch its own missiles just in case?

The loss of military spy satellites would place the world in a very precarious situation, with global nuclear war a not-unlikely outcome. Many countries consider an attack on their satellites to be an act of war. In some of the scenarios described in this book, humans may not cause the loss of satellites at all. But, in those very critical first few hours after their loss, it is unlikely that military and civilian authorities would know if their satellites were lost to an act of war or an act of God.

On a tactical level, losing satellite imagery can dramatically affect the outcome of a non-nuclear and more traditional military engagement. Consider the soldier in the modern battlefield and you will quickly understand why.

Where is the enemy? Sometimes, it is simply too dangerous for aircraft to fly over an area controlled by the enemy. (Recall what happened to Francis Gary Powers, described above.) Timely satellite data can determine their most recent location and, depending upon the speed with which the data are made available, where they are headed.

What terrain is between "us" and "them" and how can it be used to maximum advantage in the conflict? Is the bridge downstream still standing or has it been destroyed? If it is still standing, then can I send my troops across it fairly quickly? If not, then a detour will slow us down. For that matter, is the stream running deep or is it currently a dry river bed?

In the Gulf War, you may recall that the Iraqi regime used tactical ballistic missiles, equipped with non-nuclear warheads, to attack Israel and US military bases in the region. At the time, it was feared that these missiles would contain poisonous gas. Fortunately, they did not. Satellite early-warning sensors, in many cases the same as described above, were used to detect the launch of Iraqi missiles and provide early warning to their intended targets. This enabled anti-missile defenses, such as the now-famous Patriot anti-missile missile battery, to be warned and ready to intercept incoming hostile missiles. It also gave warning to the target areas and civilian population centers nearby, allowing them to take shelter and have ready access to their gas masks, should they be necessary.

With the all-too-real threat to existing orbital spy satellites, military forces around the world are seeking ways to develop "launch on demand" small satellites capable of providing near-real-time imagery of the battlefield within minutes or hours of launch – hoping they will last longer than whatever caused the existing satellites to be destroyed.

For military users, weather is critically important to the outcome of a battle. In fact, weather has historically been the deciding factor in several significant military engagements. Examples include the destruction of the Spanish Armada in 1588 as it sailed for England with the goal of dethroning Elizabeth I. The at-that-time unpredictable weather of the North Atlantic went against the Spanish when severe storms pummeled their fleet, resulting in dozens of ships being destroyed and the attempted invasion thwarted. Another and more modern example was the invasion of Normandy during the Second World War. The invasion was supposed to happen on 5 June, but the weather was poor, putting the liberation of Europe at risk. The weather cleared the next day, allowing the Allied troops to invade France and begin the liberation. Hitler, too, used the weather to his advantage, attacking the advancing Allies in the cold winter of 1944 during poor weather, knowing that the Allied foot soldiers would be busy

Figure 7.4 An American patrol crosses a field in Luxembourg during the Battle of the Bulge. (Image courtesy of the US Army)

trying to keep warm and their air forces would be blinded or grounded. The resulting conflict was known as the Battle of the Bulge (Figure 7.4).

Today, real-time satellite imagery of the battlefield can give the side with the best weather knowledge significant advantages. For example, knowing when the cloud cover will break over any given position on the battlefield will enable the best use of air power in support of whatever ground operations might be taking place. Having an understanding that your opponent is in the middle of a severe storm might give your side the extra time it needs to take a better position for a forthcoming engagement.

In addition to the weather maps and forecasts we see daily on The Weather Channel or the evening news, military weather satellites can see weather events on a much smaller scale (2–4 km or less), allowing the precise mapping of fog, sandstorms, or localized rain showers. Typically, geostationary satellites can resolve local weather events on a 10-km scale and, since they maintain their relative position over one side of the globe, their data are nearly instantaneous. The smaller-scale satellite imagery is provided by lower-flying polar-orbiting satellites that pass overhead every few hours, providing higher-resolution but not necessarily current weather data.

Satellites are also used to "listen in" on the communications of real and potential adversaries. "Echelon" sounds like the code name of a super-secret project housed in the labyrinthine bureaucracy of the US intelligence system and, in fact, that's exactly what it is. Although its specific ability to intercept global communications is not publicly known, there is enough known about Echelon to assume that most satellite, broadcast, and internet communications are monitored by spacecraft that are part of the Echelon intelligence-gathering system [4].

With codenames like Jumpseat, Trumpet, Aquacade, and Vortex, space-based signal intelligence satellites were being launched as early in the space race as their photographic cousins. To give an idea of their size and scope, the US Trumpet satellites launched in the later 2000s are rumored to weigh 5,200 kg and have antennas as large as 150 m.

To target terrorists or other enemy combatants, military planners regularly use information from intercepted radio and cell phone conversations. In Afghanistan, the signal intelligence obtained from satellites and high-altitude aircraft are instrumental in knowing where to direct drone strikes. In times of conflict, losing access to this valuable information would be devastating.

Military communications will be covered in Chapter 8.

References

[1] Graham, Col. R.H. *SR-71 Blackbird: Stories, Tales and Legends*. MBI Publishing Company, St Paul, MN (2002).
[2] Divine, R.A.; *The Sputnik Challenge: Eisenhower's Response to the Soviet Satellite*. Oxford University Press, New York, NY (1993).
[3] Callmers, W.N. (ed.). *Space Policy and Exploration*. Nova Science Publishers, Hauppauge, NY (2008).
[4] Sloan, L.D. ECHELON and the Legal Restraints on Signals Intelligence: A Need for Reevaluation. *Special Symposium Issue: Congress and the Constitution*, **50**(5) (March 2001), Duke University School of Law.

8 Communications

If we were to suddenly lose our satellites, then you could immediately say "goodbye" to:

- network television;
- cable television;
- some long-distance phone services, particularly if you are calling someone in another country;
- cell phones;
- national radio programs;
- some internet services, if not all;
- satellite phones (of course!);
- global military communications.

With the sudden loss of all the other capabilities described in this book, this is perhaps the one you would notice first. Now that we live in an information society, or an information-dependent society, depending upon your point of view, it seems we cannot go anywhere or do anything without being electronically reachable. We awaken with syndicated radio programs on our clock radios – programs that are distributed nationally to many radio stations by satellite. Then we turn on the television news or The Weather Channel to see whether we need to bring our umbrellas with us during the day and to find out who won the ballgame the night before. Cable television is really a redistribution of channels provided to your local cable provider by satellite link.

After the typical office worker arrives at work, their computer comes on. Much of what you see on the internet is carried by landlines and optical fiber, but not all – many feeds are linked through satellites. The cell phone will likely not work; the cell relay towers require GPS for timing information in order to function. And, of course, there are those who depend upon satellite phones.

Satellite Phones

Satellite phones are used by people who need to be in communication 24/7 no matter where on the globe they may be. Satellite phone handsets are slightly larger than a regular cell phone but they have the advantage of being usable virtually anywhere on the planet. All that is required is line-of-sight access to one of many satellites with transponders dedicated to providing satellite phone coverage.

The Iridium satellite phone can provide voice, paging, and low-bandwidth internet services from over 60 satellites orbiting the Earth at an altitude of 781 km [1]. The company that owns Iridium plans to launch a new network of replacement satellites with higher throughput to accommodate significantly more data (internet) traffic. In an event relevant to the subject of this book, one of the Iridium satellites was the victim of a collision in 2009 when it was hit by a dead Russian satellite, the Cosmos 2251. This event created a significant amount of orbital debris. You can read more about this collision in Chapter 1.

Globalstar is a satellite network offering voice, message, and data services to much of the globe. Their satellite constellation consists of 48 satellites orbiting the Earth at about 1,400 km. Their orbits do not permit them to have global coverage, but they are close, providing about 80% global coverage to their estimated 300,000 subscribers [2].

Finally, Thuraya is a global satellite phone company taking advantage of satellites in geostationary Earth orbit to provide voice, message, and data for much of the globe. Thuraya is estimated to have over half a million subscribers around the world.

Satellite Internet

While currently somewhat of a niche market due to the limited amount of data that can be transmitted and received, satellite internet is nonetheless in use by customers in areas not readily covered by other internet service providers such as rural and remote land areas, ships at sea, and airplanes in flight. Never underestimate the commercial potential of business executives trapped on an international flight for 8–14 hr and otherwise without the connectivity they are used to and demand.

The niche market for these services may soon be changing due to increased capability now coming online in the form of more powerful and more capable satellites dedicated to providing this sort of service.

Network Television

Network television existed before the invention of the communication satellite. First using telephone lines and then some dedicated coaxial cable connections and microwave relays to send their programming to affiliated stations, the television networks were quite adept at transmitting their programming terrestrially. But this cable and microwave relay approach would not allow them to have a global, real-time reach.

Since the launch of the Telestar satellite in 1962, the world has been using satellites in geostationary orbit to distribute television programs from a central studio to affiliate stations and cable television operators around the globe. Because these satellites are in geostationary orbit, their relative position to any ground station within line of sight (Figure 8.1) remains unchanged and they can

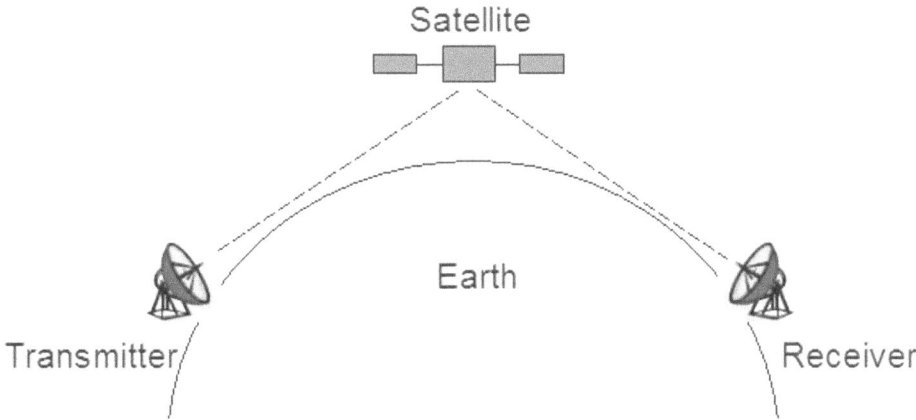

Figure 8.1 To send a signal from one side of the Earth to another, we use a communications satellite to serve as a relay. Radio signals move in a straight line and will not usually penetrate the Earth. (Image author's own)

reliably serve as a distribution point for programming from nearly all of the major television networks, including those that are unique to cable television.

The technique is simple in theory, but the application can get quite complex. The transmitting station sends the signal to the satellite, which then retransmits it down to one or more receiving stations somewhere else on the globe. As you can see, since radio travels in essentially a straight line, the mass of the Earth between the transmitting station and the receiver otherwise prohibits sending the television signal directly.

According to the Federal Communications Commission (FCC), the first credited international broadcast via satellite was a short program broadcast between a television station in Andover, MN, United States, and Pleumeur-Bodou, France. Today, all of the commercial television networks are using satellite communications to distribute programs and to provide live coverage of significant events. Instead of a single receiving station, there are many – at least one for every "local" television station in the world.

Television is a huge business in terms of revenue. NBC is purportedly charging over $3 million for an ad at the annual Super Bowl [3]. Admittedly, the Super Bowl is an extraordinary event, allowing it to command a high price for advertisers, but is simply one of many programs on thousands of stations worldwide.

Many are fond of denigrating television and the time we waste being plugged in and watching seemingly mindless programming. However, there are many important and influential programs on television that would be sorely missed should the satellite distribution network vanish:

- Real-time news coverage of events anywhere on the globe. Who didn't turn on their televisions when the Twin Towers were hit by terrorists flying airplanes on 11 September 2001? With the major economic and potential human disaster that would follow the loss of our satellite capabilities, I suspect a similar or greater number of people would go to their televisions first for news of what is going on.

- Weather forecasts and alerts. Though the loss of weather satellites is covered elsewhere (Chapter 9), the loss of national weather forecasts from sources such as The Weather Channel would be felt, quite literally, by every subscriber who depends on up-to-date weather information to plan their day or their upcoming business trip.
- Stock news. Not all of us are day traders on Wall Street, but almost all of us have an interest in what the stock market is doing – especially in the wake of whatever event precipitated the loss of our Earth-orbiting spacecraft.
- Sports. While not strictly critical to our everyday lives, many people do care very deeply about their favorite sports teams. I surmise the loss of ESPN would be the most lamented by (former) viewers!
- Entertainment. Technically a luxury, but the television entertainment industry is a huge business and the many thousands of people who rely on it for their livelihood would take exception to anyone saying its loss would not be very important.

Clearly, this is an industry that would be in crisis should it lose its primary means of program distribution and acquisition via satellite.

Satellite Television

For many, satellite television is a viable alternative to cable. Satellite television is just what it says – television signals are broadcast to your home directly from a satellite in space. There is no middleman or cable company to aggregate the many available channels and pipe them to your television set. The satellite television provider does this for you and sends the package directly to your home.

Satellite television is especially popular in rural areas, where it is often impossible to receive over-the-air television broadcasts and where cable providers may not provide service, though this satellite-provided service is available to anyone with an unobstructed view of the satellite broadcasting the signal from space. Who hasn't seen the ubiquitous home satellite dishes (Figure 8.2) in yards, on roofs, or attached to the side of apartment buildings just about anywhere on the globe?

Losing satellite capability would be a double blow to home satellite television providers who rely on two separate satellite systems to provide their programming (Figure 8.3). First of all, the network stations still provide their original programming via satellite. For the home satellite provider, they must have a dish capable of receiving these broadcasts, just like a cable television provider or a local network affiliate. They then aggregate the various programs they receive

Figure 8.2 Home satellite television antennas are small and relatively inexpensive. They are also in very wide use. (Image courtesy of DirecTV)

from multiple satellites (not all broadcasters use the same satellite to relay their programming) and broadcast the new, aggregated signal to their dedicated satellite link, which then sends the programs to the ground receivers and our homes.

Broadcast Radio

What began as a primarily local endeavor, radio soon became national and is now international in scope. Preceding television, radio provided the model for the generation, collection, and distribution of programming to a network of stations around the world. And, like television, much of this programming is

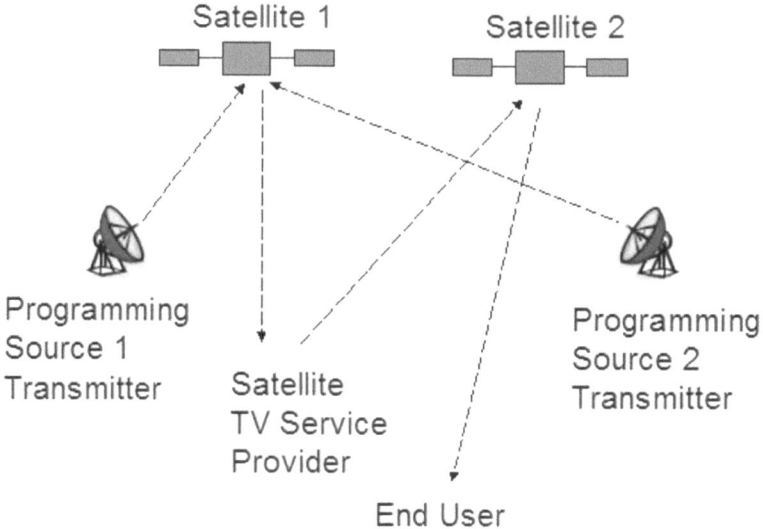

Figure 8.3 Satellite television relies on multiple satellites to deliver its programming to the end user. (Image author's own)

now distributed by satellite, cable, or fiber to your local radio station. Many radio stations get their news and programs from a major network like ABC, NBS, or CNN, while relying on other networks for their baseline programming. As a frequent radio listener (while I drive), I can reliably catch the latest news every hour, on the hour, by going to just about any broadcast station. This news feed originates from a central source and is often transmitted by satellite.

Satellite Radio

In the United States, Europe, and some other countries, direct-from-satellite radio is available on a subscription basis. In addition to paying a monthly fee, the user buys a specially designed receiver and a small antenna suited just for receiving satellite radio signals. Much of the music heard in businesses today is piped in from either satellite or another distribution channel. With hundreds of specialty channels available, many on a commercial-free basis, this convenience radio service is now big business. How big? The primary satellite radio provider in the United States is SiriusXM, who reported over 20 million subscribers in 2011.

Military Communications

Consider this: the US military, just the US military – not including the militaries of other nations – maintains $41 billion dollars' worth of space-based communications systems to coordinate operations for aircraft, ships, soldiers in the field, and other sites around the globe. This includes a network of satellites such as Milstar (Military Strategic and Tactical Relay) and the Defense Satellite Communications System (DSCS). There are many more satellites and satellite constellations in the military communications network than can be described here and all of them are vulnerable [4].

Milstar was designed to be used during war to provide secure and survivable communications. The system and spacecraft were designed to be resistant to jamming and the space nuclear effects that might occur during an all-out war. The system is designed to allow the control of submarines and submarine-launched missiles, aircraft, and soldiers on the ground.

Adjunct to Milstar is the DSCS, which ties the armed forces of the United States closely with their command and control in the United States, including the White House and the US State Department. The DSCS consists of eight satellites in geosynchronous orbit and others.

Quoting from the US Army's Tactical Satellite Communications Field Manual (Number 24-11):

> "a. Command, control, and communications (C3) is the key to success in the Air Land Battle. Due to technological advances, greater mobility, and the extended battlefield, radio communications is paramount in the communications plan. However, while technology has improved the equipment, communications has not kept pace. Two limitations are the congested frequency spectrum and the physical limits on radio wave propagation. The frequency required for long-range radio adds to the frequency congestion problem. Requirements normally exceed the available, useable frequencies. Frequency congestion and inherent limitations of terrain and noise hamper short-range tactical radio. Coupled with the need for flexibility, security, and reliability, radio communications remains a critical problem to the communicator.
>
> b. TACSAT communications is the first radio system to successfully overcome most of these limitations. Using an orbiting satellite repeater illuminates one-third of the earth for direct line of sight (LOS) operations. This makes it possible to establish tactical communications on a scale never before accomplished. With more frequencies available and a single station LOS relay to almost any point on the battlefield, TACSAT equipment greatly enhances communications."

Given the above publicly acknowledged dependence on space satellites for battlefield communications, what would be the result if these satellite resources were denied to soldiers trained to depend upon them?

Soldiers are also being equipped with smart phones and applications tailored to their use in the battlefield. These applications, which will use the phones' built-in GPS capability and access to cellular networks, may make the soldier more vulnerable than ever before if the satellites upon which both of these services depend become unavailable. For the modern soldier, raised on nearly instant communication with just about anyone, anywhere on the globe using a smart phone in the palm of his or her hand, losing this capability will be akin to fighting without a weapon.

References

[1] Kota, S.L. Broadband Satellite Networks: Trends and Challenges. IEEE Wireless Communications and Networking Conference, Sunnyvale, CA, March 2005.
[2] Kumar, S. Mobile Communications: Global Trends in the 21st Century. *International Journal of Mobile Communications*, **2**(1) (2004).
[3] Cost of Average Super Bowl Commercial? $3.5M. USA Today online, *http://usatoday30.usatoday.com/sports/football/nfl/story/2012-01-03/super-bowl-ad/52360232/1*.
[4] Elfers, G.; Miller, S.B.; Future US Military Satellite Communications Systems. The Aerospace Corporation, available online at *www.apcon.aero/news/gapfill_future.pdf*.

9 Weather Forecasting

On 8 September 1900, the residents of the island town of Galveston, Texas, knew there was a storm coming. What they didn't know was that the storm was a Category 4 hurricane and that, before the next day was through, over 6,000 residents of this tranquil Texas town would be dead. The hurricane's storm surge was over 15 feet and the highest point on the island was under 9 feet – meaning that the entire island was under water at the peak of the storm's fury [1].

Thanks to modern weather-forecasting technology, which includes global satellite imagery, my family was able to evacuate the Alabama Gulf Coast town of Gulf Shores three days before Hurricane Ivan made landfall in 2004. Ivan, a Category 3 storm when it struck Gulf Shores, had been tracked since it formed in the mid-Atlantic and strengthened over the warm waters of the Gulf of Mexico. At one point, it was a Category 5 storm. We had just enjoyed a week of sunny days on the beach when the hurricane warning signs began to go up. The sky was clear and hundreds of thousands of Gulf Coast residents were on alert, watching The Weather Channel as Ivan closed in.

Figure 9.1 Hurricane Ivan as seen from the International Space Station in 2004 just before it made landfall on the US Gulf Coast. (Image courtesy of NASA)

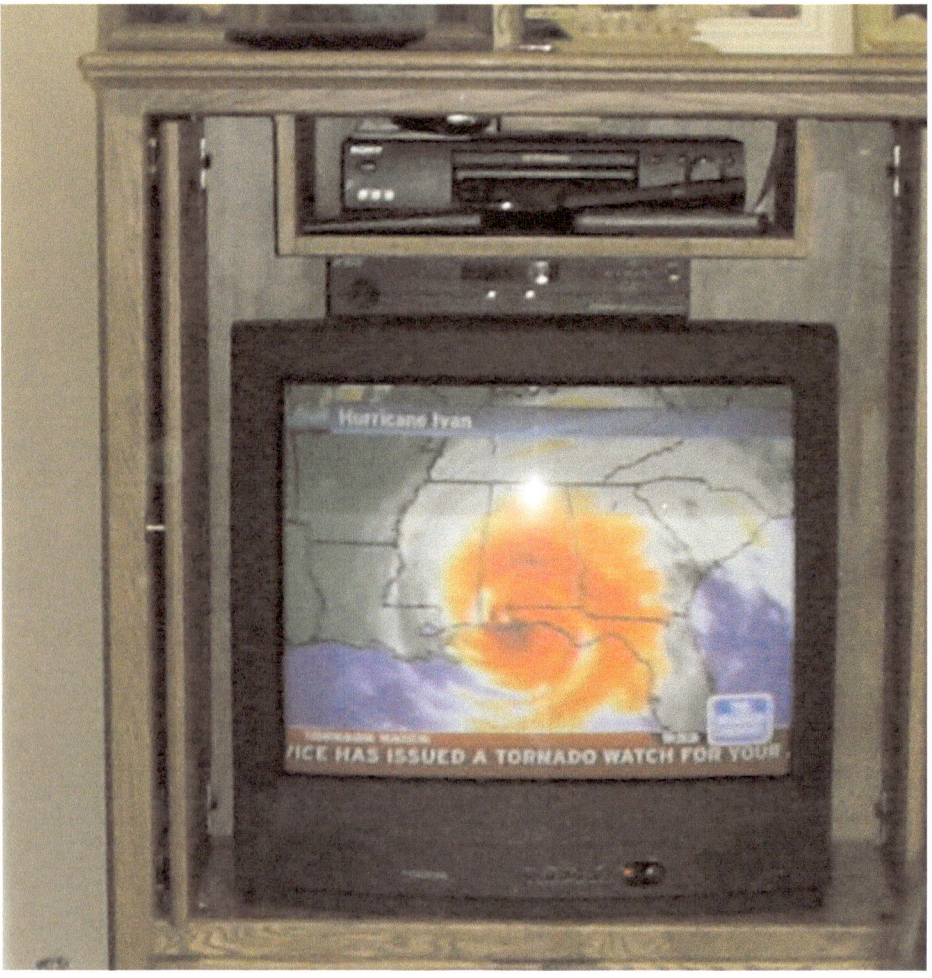

Figure 9.2 Hurricane Ivan making landfall near Gulf Shores Alabama as seen from our living room television nearly 300 miles inland – at the top of Alabama – where we were experiencing heavy rain and strong winds from this monster storm. (Image author's own)

When we left Gulf Shores, the sun was shining and there was no physical evidence that a hurricane was imminent. Yet, as can be seen in Figure 9.1, residents of the International Space Station (ISS) knew that Ivan was a monster and that it was headed our way. Like so many others, we heeded the warnings that were unavailable to the residents of Galveston in 1900 and departed for home – some 300 miles inland.

Two days later, Figure 9.2 shows what we watched on our television as the storm devastated the beach community in which we had been vacationing just days before.

Ivan hammered the Gulf Coast and few, if any, lost their lives, despite the fact that the coast is much more populated today than it was in 1900. Satellite imagery and weather forecasts make accurate predictions like this possible, saving lives and property.

Other than the pretty pictures taken aboard the ISS, how does this "storm story" relate to space satellites?

Several countries have weather satellites in space and many collaboratively share data to assure global, uninterrupted weather forecasting. Defined by their orbits, there are two types of weather satellites: (i) those than look at one hemisphere of the Earth continuously from geostationary Earth orbit and (ii) those that fly over a given piece of land area about once every 12 hr from a lower-altitude, polar-orbiting satellite.

Within these two categories, there are weather satellites with cameras that look at the Earth in visible light, making observations of hurricanes and other weather systems relatively easy and whose images would, at least on a first-glance level, be understandable to just about anyone who sees them. Some satellites are equipped with cameras that see heat instead of visible light. These heat-observing (or infrared/IR) cameras are useful for determining the height of the clouds that are perhaps visible to the naked eye, as well as ocean and land surface temperatures (Figure 9.3). Information such as this goes into complex weather-prediction computer models which produce the remarkably accurate (but still not perfect) weather forecasts of today.

The usefulness of the data collected by these satellites depends upon their resolution (what is the smallest object or weather event they can see?) and the amount of time between two consecutive images (how fast is the storm moving and in what direction?). Typical weather-image resolution made available to the public is in the order of 10 km. These images come from the satellites in geostationary Earth orbit (GEO). For closer-in views, users rely on the satellites orbiting the Earth's poles. While an individual satellite may only provide two images per day, they are often showing details with a resolution of 2 km or less. (It has been reported that some defense meteorological satellites can provide a resolution smaller than one half of 1 km.) These higher-resolution images are much more useful in seeing things like tornados and sandstorms, which are small compared to hurricanes, but nonetheless destructive.

Today, over 100 million people live along the Atlantic and Gulf Coasts, potentially in the path of (future) killer storms like Ivan, Katrina, and others. If we were plunged back to a time without satellite imagery, we would still be able to track hurricanes, but with much less accuracy and much more risk. Aircraft patrolling the oceans would undoubtedly see the storms coming and probably make fairly good estimates of their strength and direction. But how well would that infrastructure hold up without the satellite assets upon which they depend, such as GPS (Chapters 4, 5, and 6) and communications satellites to relay their radio messages (Chapter 8)?

Let's move away from massive storms and talk about the day-to-day and more localized weather forecasts. Would they be affected if satellite data were to be lost?

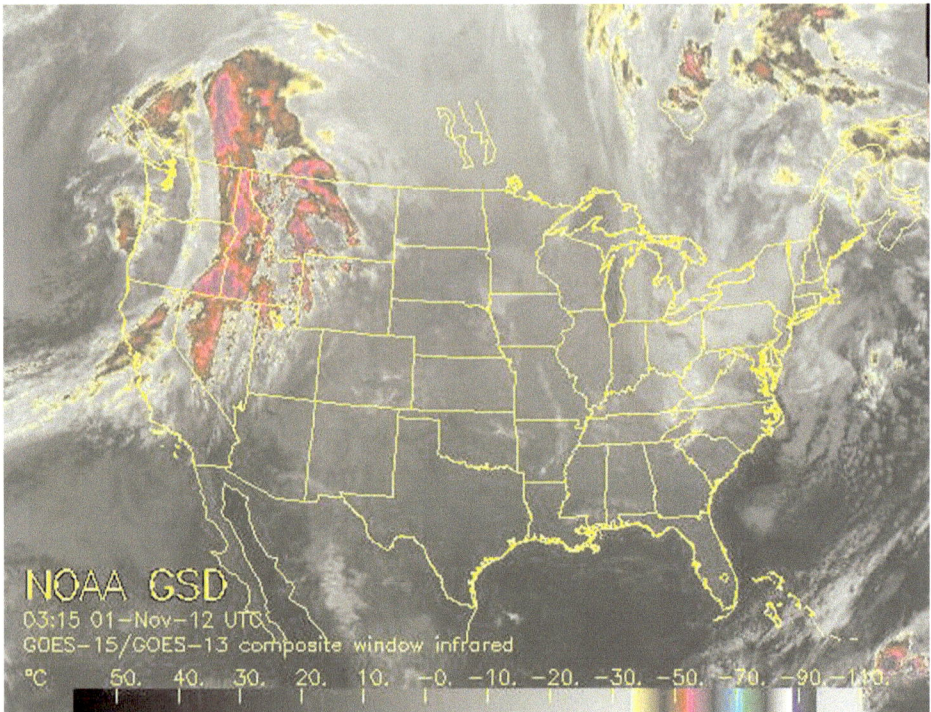

Figure 9.3 This infrared (which means it shows relative temperatures) image of the United States was taken by the Geostationary Earth Observing System (GOES) satellite in 2012. Seen in the image are clouds; warmer clouds are darker than cooler ones. (Image courtesy of the National Oceanic and Atmospheric Administration)

Weather forecasting is, to put it mildly, complicated. Sophisticated computer models are required to take into account the huge amount of data that goes into making an accurate weather forecast. Much of this comes from ground observations and instruments, and is combined with data from aircraft and space satellites to provide a comprehensive view of both local and regional weather.

Even without a global space satellite disaster, our weather prediction and warning capabilities are at risk. The National Research Council (NRC) issued a report in April 2012 warning that our system of weather and Earth-observing satellites is "beginning a rapid decline" [2]. The reason? We haven't replaced enough of the currently operating satellites fast enough to replace those that are failing. The NRC projects that we might lose greater than 70% of this type of satellite by 2020, leaving a gaping hole in our observational capabilities.

It is important for the ground observations to be from known locations (GPS!), appropriately correlated in time (GPS again!) and reported in a timely manner (ground and cellular networks – also affected in performance by GPS as described in Chapter 8). If we lose these supporting infrastructures, the impact on commerce, agriculture, and everyday life would be widespread and profound:

- Severe weather advisories, alerts and warnings would be much less accurate and their geographic specificity would suffer. Instead of being able to predict with high accuracy which specific areas of a state or county are at risk from a severe storm, warnings would become more regional and less timely.
- Air traffic control for civilian aircraft would be hit particularly hard. Aside from the loss of navigation information with the GPS satellites down (Chapter 5), air traffic controllers would have less information about weather systems, requiring aircraft re-routing as well as localized weather take-off and landing delays.
- Farmers rely on long-range weather forecasts for timing the planting of crops and their harvesting.
- Ships at sea and on inland waterways would have to revert to pre-satellite methods of acquiring weather information.

Reference

[1] Bixel, P.B.; Turner, E.H. Galveston and the 1900 Storm. University of Texas Press, Austin, TX (2000).
[2] *www.nationalacademies.org/nrc/index.html.*

10 Remote Sensing: Environmental Monitoring and Science

Energy, environment, farming, mining, land use. All of these areas and more are now inextricably linked to satellite data and would be devastated should that flow of data stop.

Environmental Monitoring

Oh how complacent we've become. We take for granted that we will have instant images from space showing a volcanic eruption somewhere in the South Pacific within hours of learning that it happened. When the BP oil spill happened in the Gulf of Mexico in 2010, satellite images were used in conjunction with aircraft and ships to monitor the extent and evolving nature of the spill (Figures 10.1 and 10.2).

Figure 10.1 Close-up view of the Gulf of Mexico oil spill as seen by NASA's Earth Observing-1 satellite. (Image courtesy of NASA)

Figure 10.2 Wide-area view showing the Gulf of Mexico and the growing slick from the subsurface oil spill. (Image courtesy of NASA's Aqua satellite)

The data were also used to direct the ships that were attempting to clean up the spill, to warn fishermen of areas in which it would be dangerous to fish, and to generally monitor the extent of the disaster. This is the type of data we get from space in a field known as remote sensing.

Remote sensing is, well, exactly what its name implies. With it, you gather data, or sense, usually in the form of electromagnetic radiation (light), remotely – that is, you are not physically touching what you are looking at. Satellite remote sensing began shortly after we began launching satellites and many industries are now totally dependent upon having the capability.

We use satellites, like the venerable Landsat series, to study the Earth in unprecedented detail. Since 1972, Landsat satellites have taken millions of high-resolution images of the Earth's surface, allowing comprehensive studies of how the land has changed due to human intervention (deforestation, agriculture, settlement, etc.) and natural processes (desertification, floods, etc.).

The best way to understand how useful Landsat and similar data can be to governments at all levels is best illustrated by looking at "then and now" photographs. For example, Africa's Lake Chad has been shrinking for 40 years, as the desert has encroached on this once plentiful inland freshwater lake. Forty years ago, there were about 15,000 square miles of water within the lake. Now, it is less than 500 square miles (Figure 10.3) [1].

And what is the practical side of this particular bit of information?

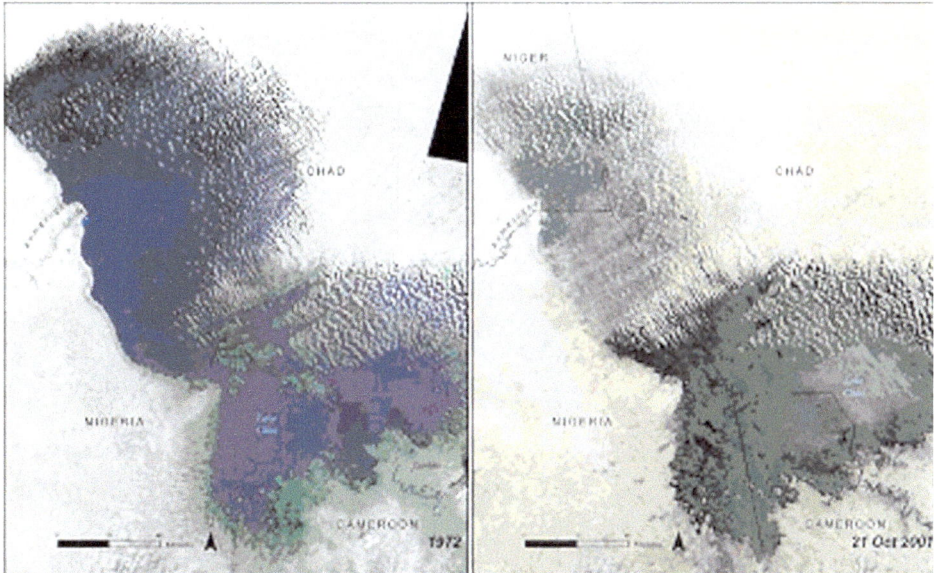

Figure 10.3   Satellite images have captured the shrinking of Lake Chad over the last few decades. The image on the left, taken in 1972, shows a large Lake Chad. The image on the right shows the lake as it appeared in 2001 – with mostly marshland remaining from what had once been deep water. (Images courtesy of NASA)

Governments use this type of satellite imagery to avoid human tragedy. Hundreds of thousands of people, if not millions, depend upon the waters of Lake Chad for agriculture, industry, and personal hygiene. With the lake going dry, how has this impacted on their livelihoods, their families, and their very lives?

The European Space Agency (ESA) is freely providing satellite data to developing countries as they search for new sources of drinking water. For example, ESA assessed data obtained from space over Nigeria to find over 90 new freshwater sources within that country. After ground teams visited the new sites, all were confirmed to contain fresh water. This was no accident. These were satellites with sensors developed for just such purposes in mind [2].

Desertification is but one example of changing climates affecting people's everyday lives. What about more direct observations of our impact on the planet? Figures 10.4 and 10.5 show the scarring of the Earth's surface as a result of surface mining in West Virginia. This is not a polemic against mining; rather, it is an observation that we can use satellite imagery to monitor such mining and be mindful of its impact on the environment.

Other than taking pictures of surface features, like lakes and open pit mines, how are satellites monitoring the Earth's changing climate? In just about every way, by: monitoring global land, sea, and atmospheric temperatures; measuring yearly average rainfall amounts just about everywhere on the globe; measuring

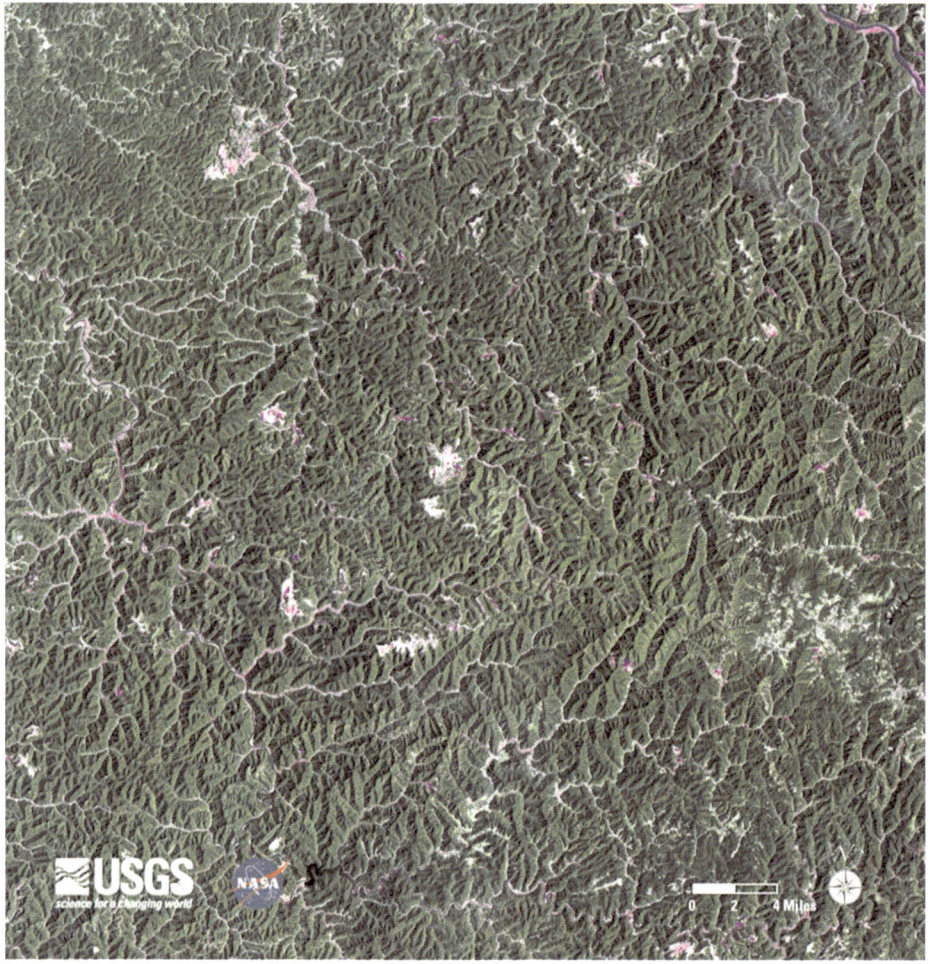

Figure 10.4 Landsat image of West Virginia showing the scarring of the Earth's surface from open pit mining in 1987. (Image courtesy of the United States Geological Survey and NASA)

glaciation rates; measuring sea surface heights; and more. Remote sensing is more than taking pictures of the Earth in the visible part of the spectrum. We can learn a great deal from looking at part of the spectrum that our eyes cannot see – but our instruments can.

Shown in Figure 10.6 is a composite image of the Earth's surface showing the average land-surface temperature at night. The data came from two NASA satellites, Terra and Aqua, as they orbit the Earth in a polar orbit. (This means that they circle the Earth from top to bottom, passing over both the North and South Poles with each complete orbit.) Terra's orbit is such that it passes from the north to the south across the equator in the morning; Aqua passes south to north

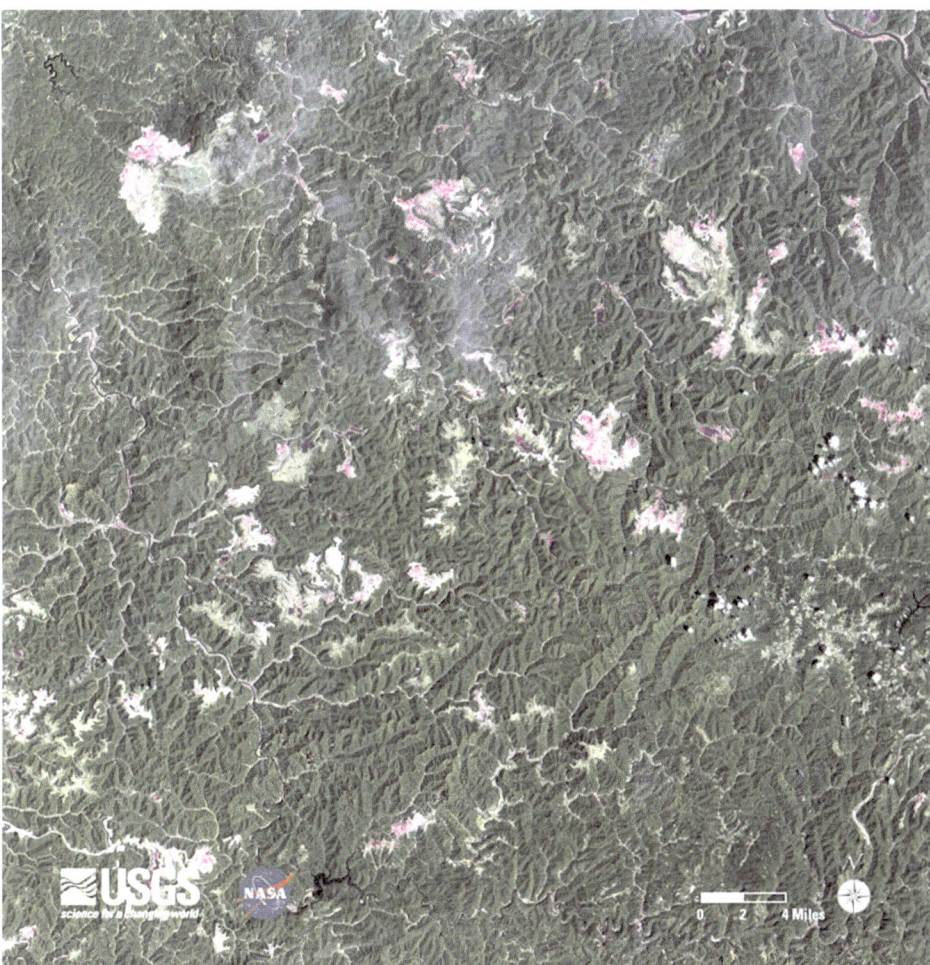

Figure 10.5 Landsat image of the same area in West Virginia in 2011 showing how much more land has been mined. (Image courtesy of the United States Geological Survey and NASA)

over the equator in the afternoon. Taken together, they observe the Earth's surface in its entirety every two days. Data sets such as this exist for just about any day of the year and can show either night-time lows or daytime highs.

By looking in different parts of the spectrum, like the infrared light discussed above, we can make observations as described in Table 10.1.

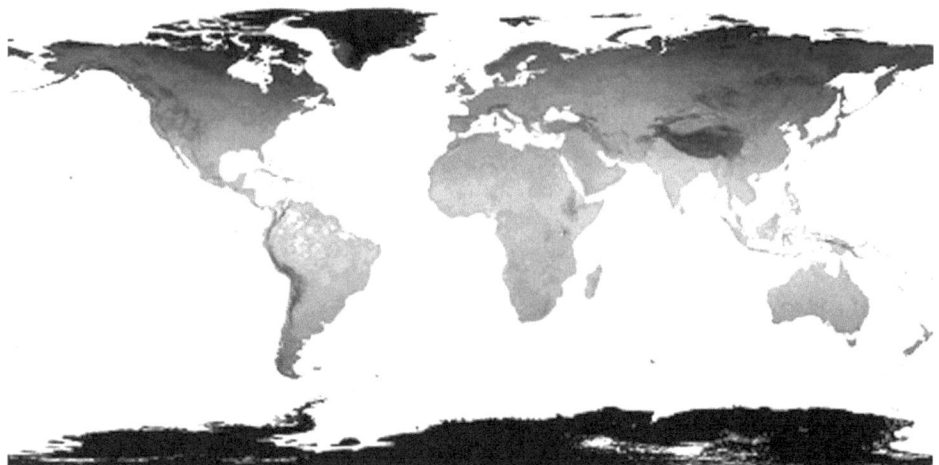

Figure 10.6 Land-surface temperatures at night, averaged over one month, as determined by NASA's Moderate Resolution Imaging Spectroradiometer instrument aboard the Terra and Aqua satellites. Darker colors are colder, lighter are warmer. (Image courtesy of NASA)

Table 10.1. Our "vision" is expanded by using infrared imaging.

Portion of the electromagnetic spectrum	What we can see
Near-infrared	Especially suited to observe the health of leafy plants
Mid-infrared	Used to measure the moisture content in the vegetation and the nearby soil
Thermal infrared	For measuring land-surface temperatures and to identify some types of rocks

Pollution Monitoring

As emerging countries industrialize, they also become polluters. Many of these countries are not exactly forthright about releasing air-pollution details to the media, so much of our awareness of the rising pollution there is anecdotal – typically in the form of stories told by people who have visited these countries and seen the extreme pollution at first hand. This, by the way, is not exactly scientific.

Using satellites, and not relying on either the governments in question or second-hand stories, we can accurately assess the pollution levels there and elsewhere. Using satellite images to measure the amount of light absorbed or blocked by fine particulates in the atmosphere, otherwise known as air pollution, you can determine not only what the airborne pollutant might be, but also its size. And, by looking at the overall light blockage, an accurate estimate of the amount of pollution in the air can also be made. Recent studies show that many of these countries are covered in a pollution cloud that countries in the developed world would deem extremely harmful. And how do we know this with scientific certainty? From satellite measurements.

Energy Production

The recent boom in the production of shale oil in the United States and elsewhere is due in large part to the identification and geolocation of promising geologic formations for test drilling and fracking. "Fracking" is a somewhat new term that comes from the phrase "hydraulic fracturing". In fracking, massive amounts of previously unusable reservoirs of oil and natural gas are released for capture, sale, and transport from deposits deep within the Earth – many located at least a mile below the surface. In the United States alone, there may be as much as 750 trillion cubic feet of natural gas within shale deposits releasable by fracking [3]. How do energy companies know where to look for these deposits? In large part, by analyzing satellite imagery.

According to *Science Daily* (26 February 2009), a new map of the Earth's gravitational field based on satellite measurements makes it much less resource-intensive to find new oil deposits. The map will be particularly useful as the ice melts in the oil-rich Arctic regions. The easy-to-find oilfields have already been found. To fuel the growing world economy, those harder-to-find deposits must be located and tapped – which is why satellite imagery is so important. Take away this and other satellite-dependent techniques of oil and gas exploration and the world economy will feel the impact through higher oil and natural gas prices.

Resource Location

Looking for rare minerals to be mined for our many gadgets, household appliances, and industrial machines? Soil type is often a strong indicator of whether or not underground deposits of metals and minerals are located. By using satellite data to identify promising surface structural features and different soil types, mining companies can better identify promising mining locations, wasting less time and effort in finding the best places to obtain much-needed industrial resources. Without satellite images, the finding and assessment of promising new mines would grind to a halt as the industries retooled back into the days of much slower and labor-intensive field surveys (but without GPS!).

Agriculture

To feed the Earth's growing population affordably, farming has gone from a mostly decentralized, family-owned business to corporate farming on a scale never before imagined. These industrial megafarms are a primary reason that many people in the world can enjoy plentiful and varied foods at a reasonable cost. On this scale, deciding what crop to plant in a given field is not just business – it's science. And the science relies, in large part, on data from space.

Companies such as the Satellite Imaging Corporation (SIC) provide data from space on overall crop health, soil analysis, and irrigation impacts and efficiencies. From space, you can easily map soil variations, finding areas rich in organic matter and others less so – this allows optimized planting to take advantage of crops that thrive in any given soil environment. Very large farms also use satellite images to assess the overall health of their crops by land area, spotting those that are being impacted by non-optimal soil moisture content, etc., allowing the farmer to take corrective action while there is still time to save the crop.

Satellite Archeology

With the staggering amount of Earth science data coming from space, very practical uses of them are quickly taking shape. For example, developing countries in Central America and Africa are beginning to use some of these "free" data to address developmental challenges (deforestation, fires, etc.) and to provide information allowing better coordination of aid following natural disasters such as hurricanes, earthquakes, and mudslides [4].

Satellite technology is now being used to aid in uncovering our past. Laser radar and altimeters and synthetic aperture radar technology flown in space are being used in conjunction with traditional archeological techniques to find lost cities and other areas of historical interest. Here are just a few headlines that will give you a feel for how valuable this tool has become:

> *"Ancient Geoglyphs Discovered Near Titicaca, Peru*
> According to an Italian scientist, a huge network of earthworks, or geoglyphs, is visible in satellite imagery of a large area, over 463 square miles, in the surroundings of the Titicaca Lake, Peru. Amelia Carolina Sparavigna, professor at Italy's Politecnico di Torino, claims the patterns she discovered while studying satellite pictures near the Titicaca Lake. She says the shapes are the result of an almost unimaginable agricultural effort of Andean communities centuries ago."
>
> *Peru This Week,* 17 October 2010

> *"Satellite Imagery Solves a 400-Year-Old Mystery*
> Bangalore: Satellite photos of Talakad, an ancient city located on the banks of the Cauvery, near Mysore, have found several man-made

canals which, archaeologists say, lend weight to the famous curse that brought this temple destination down.... The research, which was conducted by the National Institute of Advanced Studies (NIAS) in collaboration with the state archaeology department, Karnataka, found a well-developed canal system extending a few kilometers from Talakad to Cauvery. "We analysed the site through geospatial maps recorded by a satellite using infrared and radar technology," said MB Rajani, the project head. "A GPS survey was done on the site for more accuracy. By analysing data and comparing it with historical evidence, we were able to arrive at the findings."

Daily News and Analysis, India, 9 June 2009

"Ancient Canals Discovered in Heart of US City
Anthropologists, with the assistance of satellite imagery, have discovered the remains of a series of ancient canals, located just south of the Salt River, near the very heart of downtown Mesa, Arizona. The existence of the canal system, built in the Salt River valley centuries ago by the Hohokam, has long been known, but the extent of this most recent discovery has caught some experts by surprise."

Digital Journal, 27 January 2009

"Ancient Peru Pyramid Spotted by Satellite
A new remote sensing technology has peeled away layers of mud and rock near Peru's Cahuachi desert to reveal an ancient adobe pyramid, Italian researchers announced on Friday at a satellite imagery conference in Rome."

Discovery News, 16 November 2008

No satellites means no insight into how humans are affecting the climate; the water cycle; sea, air, and land temperatures; and the creation and spread of air pollution. No satellites also means a decreased ability to monitor crops for disease, predict landslides and mudslides from soil moisture measurements, and degraded overall land management.

NEWSFLASH: On 18 April 2012, the Landsat 7 satellite had to maneuver to avoid colliding with a piece of space debris.

References

[1]   Coe, M.T.; Foley, J.A. Human and Natural Impacts on the Water Resources of the Lake Chad Basin. *Journal of Geophysical Research*, **106**(D4) (2001).
[2]   From the European Commission Enterprise and Industry website, *http://*

ec.europa.eu/enterprise/policies/space/esp/international-cooperation/africa/index_en.htm.

[3] *Fueling North America's Energy Future, Executive Summary.* IHA CERA Special Report, Cambridge, MA (2010).

[4] Hardin, D.M.; Irwin, D.; Sever, T.; Graves, S. Realizing NASA's Goal of Societal Benefits from Earth Observations in Mesoamerica through the SERVIR Project. American Geophysical Union, Fall Meeting 2006.

11 The International Space Station and Human Space Flight

The loss of a satellite due to a collision with orbital debris is one thing – an expensive event for sure, but, if it is a rare event, it can be overcome. The loss of a human life in a collision is quite another thing indeed.

In June 2011, the astronauts living in the International Space Station (ISS) had to board the station's lifeboats after an errant piece of orbital debris was spotted and found to be heading their way [1]. The astronauts left their duty stations and boarded the two Russian Soyuz capsules that serve as their escape pods (allowing them to return to the Earth at any time in the event of an emergency) until the threat was over. This was one of the few times in the history of the ISS that the crew was forced to take this action in response to a possible space collision. Under normal circumstances, potential collisions are identified well in advance, allowing the ISS to use on-board propellant to move to a slightly different orbit, thus reducing the chance of collision. Not so this time – the debris object was detected too late to allow the ISS to successfully maneuver.

As large as an American football field (Figure 11.1), the ISS makes a large, expensive, and quite vulnerable target for orbital debris.

As discussed previously, NASA and other countries monitor the orbital debris environment in space, tracking and predicting the trajectories of objects as small as 5 cm in diameter. The data are then assessed to determine collision risks with other space satellites and objects, including the ISS. NASA and its international partners have a set of guidelines that are used to determine whether the station has to take some sort of evasive maneuver or not. The process for doing this is best described by NASA on its website devoted to orbital debris issues as they relate to the ISS:

> "These guidelines essentially draw an imaginary box, known as the 'pizza box' because of its flat, rectangular shape, around the space vehicle. This box is about a mile deep by 30 miles across by 30 miles long (1.5 × 50 × 50 kilometers), with the vehicle in the center. When predictions indicate that the debris will pass close enough for concern and the quality of the tracking data is deemed sufficiently accurate, Mission Control centers in Houston and Moscow work together to develop a prudent course of action.
>
> Sometimes these encounters are known well in advance and there is time to move the station slightly, known as a 'debris avoidance

Figure 11.1 The International Space Station (ISS) is over 72 m long, making it a large target for orbital debris. (Image courtesy of NASA)

maneuver' to keep the debris outside of the box. Other times, the tracking data isn't precise enough to warrant such a maneuver or the close pass isn't identified in time to make the maneuver. In those cases, the control centers may agree that the best course of action is to move the crew into the Soyuz spacecraft that are used to transport humans to and from the station. This allows enough time to isolate those spaceships from the station by closing hatches in the event of a damaging collision. The crew would be able to leave the station if the collision caused a loss of pressure in the life-supporting module or damaged critical components. The Soyuz act as lifeboats for crew members in the event of an emergency.

Mission Control also has the option of taking additional precautions, such as closing hatches between some of the station's modules, if the likelihood of a collision is great enough.''

(*www.nasa.gov/mission_pages/station/news/orbital_debris.html*)

NASA will conduct the debris-avoidance maneuver when the probability of collision is greater than 1 in 100,000 if the maneuver doesn't cause a significant disruption in the mission's current objectives. If the risk is determined to be greater than 1 in 10,000, then the maneuver will occur regardless of its scientific impact [2]. Due to the complexity of the ISS and the coordination required among the various international partners, conducting an avoidance maneuver requires a little more than one day's notice.

The collision risk between debris and the ISS is increasing.

In March 2012, the ISS crew had to again take shelter when a piece of a Russian Cosmos 2251 satellite passed within 14 km of the station [3]. Recall that these objects are not creeping along or simply "floating by". They are traveling at orbital speeds and, depending upon the direction from which they come relative to the direction in which the ISS is moving, with a relative collision velocity of twice their orbital velocities.

In the first 12 years on orbit, the ISS had to take evasive action to avoid a collision with debris only about once per year. That all changed in 2011. In the period between April 2011 and April 2012, the ISS took evasive action four times [4]. Table 11.1 very clearly shows the relationship of the increased collision risk to the ISS with two relatively recent events: the Iridium/Cosmos collision and the Chinese anti-satellite test that destroyed their Fengyun-1C weather satellite (Chapter 1).

Table 11.1. The International Space Station's dramatic increase in "close calls" with orbital debris, beginning in early 2011.

Date of ISS close approach	Debris object	ISS action taken
April 2011	Piece from Russian Cosmos satellite	Evasive collision-avoidance maneuver
June 2011	Piece of Russian Proton rocket motor	Station crew entered escape vehicles for possible evacuation
September 2011	Russian rocket body part	Evasive collision-avoidance maneuver
January 2012	Piece of American Iridium-33 satellite	Evasive collision-avoidance maneuver
January 2012	Piece of Chinese Fengyun-1C weather satellite	Evasive collision-avoidance maneuver
March 2012	Piece from Russian Cosmos satellite	Station crew entered escape vehicles for possible evacuation

The ISS has been hit by small pieces of debris several times. Thankfully, none of the strikes so far has caused life-threatening damage. In June 2012, the station's cupola, a glass "bay window" that allows the crew to have stunning views of the Earth and deep space from within the station, was hit by a small piece of debris [5]. Wires connected to the station's solar arrays have also been damaged by debris strikes. It is only a matter of time before a more serious event occurs.

The ISS is not the only human spacecraft to be so threatened. The Chinese are now routinely sending people into space as they prepare to build their own space station by the middle of this decade. It is ironic that the not-yet-flown space station will have to be designed to survive or avoid collision with the huge amount of debris generated by their own anti-satellite weapons test in 2007 (Chapter 1).

Private space companies are developing the capability to send tourists into space for extended periods of time; some are even planning to place hotels in orbit. Bigelow Aerospace plans to launch its BA-330 Space Station, which has the advertised capability to house up to six people for an extended period of time. If they are successful, then they, too, will have to take into account the risks posed by debris to their crews and establish their own "abandon ship" criteria.

The ISS and human exploration craft are vulnerable to solar storms as well. Any storm strong enough to knock out our military and civilian satellites will certainly threaten crewed systems, from the impact not only, as one might suspect, to the sensitive electronics that control the affected spacecraft, but also to their human crew. Both the primary radiation from the storm and the secondary radiation it can create as it passes through and interacts with the thin metal skin of the ISS can pose a significant health risk. The crew of the ISS has had to take shelter in more heavily shielded areas of the ISS several times in its history thanks to variations in the Sun's storm activity.

The particles emitted by the Sun during a solar storm, typically high-energy protons and sometimes other elements, are all considered to be ionizing radiation. The energy that ionizing radiation loses as it travels through a material or living tissue is absorbed by that material or living tissue. The ionization of water and other cell components can damage DNA molecules near the path the particle takes, a direct effect of which is breaks in DNA strands including clusters of breaks near one another – breaks that are not easily repaired by cells. Such DNA break clusters are much less frequent, or do not occur at all, when cells are exposed to the types of radiation found on the Earth. Because it can disrupt an atom, space radiation also can produce more particles, including neutrons, when it strikes a spacecraft or an astronaut inside a spacecraft – this is a secondary effect.

Health risks from radiation exposure may be acute, immediate effects or longer-term effects. The extent and severity of acute effects are determined by the type and amount of radiation exposure, and they range from mild and recoverable effects, such as nausea and vomiting, to central nervous system damage and even death. Longer-term effects include increased cancer risk.

Exposure to radiation may cause ionizations of the molecules of living cells. At low doses, such as that which we receive every day from background radiation, cells can repair the damage rapidly. At higher doses, the cells may not be able to repair themselves and they can either be changed permanently or die. It is the latter that is of most concern if there is a large solar storm.

The mechanism that causes cell damage or death is actually rather simple. The ionizing radiation collides with a DNA molecule, causing it to break. A DNA molecule carries the genetic information that makes us who we are. It is composed of two strands, linked together to form bonds with each other, and looks like a twisted ladder (Figure 11.2). A break at either end of the strand is relatively easy for the body to repair, but breaks to both strands are much more difficult to repair, and the affected cells may either be changed permanently or die.

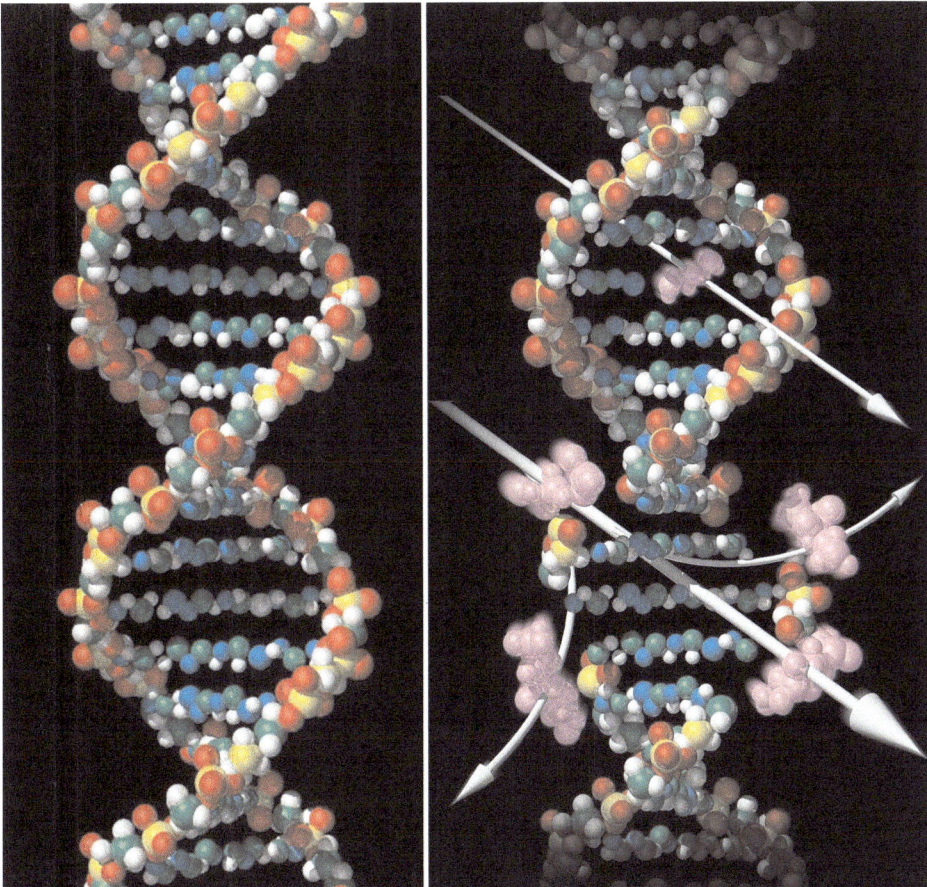

Figure 11.2 Ionizing radiation can damage DNA, leading to an increased risk of cancer and, if the damage is severe enough, radiation sickness and death. The left side of the image shows undamaged DNA. The right side shows DNA after being struck by ionizing radiation. (Image courtesy of NASA)

In the event of a very large storm, the crew is at risk of immediate death from the radiation they might receive. On 20 January 2005, the Sun unleashed a massive storm that resulted in near-Earth space not protected by the Earth's magnetic field experiencing radiation levels previously unheard of [6]. If the Apollo astronauts had been walking on the Moon at the time, then they would have received a very large dose of radiation in a very short amount of time. According to Francis Cucinotta of NASA's Johnson Space Center, "An astronaut caught outside (on the Moon) when the storm hit would've gotten sick" [7]. The astronaut would, within only a couple of days, begin to experience vomiting, fatigue, and other symptoms of radiation sickness; thankfully, the dose would probably not have been fatal.

The same cannot be said for the flare of August 1972. Just months before the launch of Apollo 17, a massive solar storm erupted [8]. It was estimated that the radiation sent into space by that storm would have been sufficient to make the astronauts extremely sick – or even kill them – had they launched earlier and been on their way to or from the Moon at the time.

Despite the risks, NASA and other countries are still planning to send humans beyond the protective shield of the Earth's magnetosphere and into deep space. The spacecraft concepts being considered for these missions would not try to shield the entire spacecraft from solar storms; they would instead have located within them a storm shelter, probably surrounded by a relatively heavy layer of proton-absorbing material.

Materials with high hydrogen content generally have greater shielding effectiveness, but they typically don't perform well as the primary structure for a spaceship. Liquid hydrogen and methane, which are often used as rocket propellant, are also good choices for radiation shields. Some plastics appear to be good shields, but it is far more likely that good old water will be used.

Within a deep-space spacecraft, the storm shelter will likely be located where the astronauts sleep, since the water shield will also provide protection from the constant flux of radiation in the solar wind and from galactic cosmic rays (ionizing radiation that comes into our solar system from elsewhere in the galaxy). During a storm, the crew could take refuge here until the immediate danger passes, which is typically in the order of hours or, at most, days. Figure 11.3 shows a recent concept proposed for a deep-space habitat. On the left are the crew quarters, around which is a bladder filled with water that provides a modest amount of shielding from solar radiation and cosmic rays.

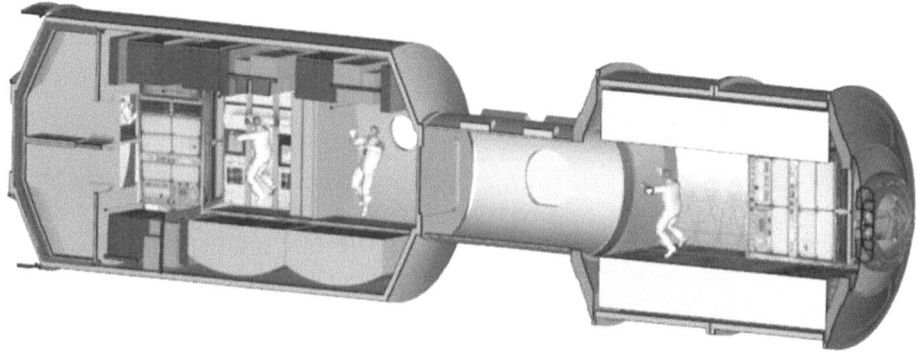

Figure 11.3 Artist concept of a deep-space habitat. Visible on the left are the crew quarters surrounded by a water shield to protect them against ionizing radiation. (Image courtesy of NASA)

Shielding a crewed spacecraft from this radiation is not an easy task. True, with enough mass, just about any cosmic ray or solar wind atom can be stopped. But mass is, by definition, heavy and therefore very expensive to launch from the Earth. Rockets are typically limited in the weight they can launch and taking enough shielding to completely protect the human crew from all radiation may well be impossible – and it is certainly impractical.

Protecting astronauts on the surface of another planet requires a different strategy. Rather than transporting the mass, and weight, of a water shield from space down to the surface, why not use part of the planet or the Moon to provide the shielding?

On the Moon, for example, astronauts might bring equipment to scrape up the relatively thin lunar dirt, called regolith, and pile it on top of the crew habitat. A few meters of dirt would be more than adequate to absorb the most lethal cosmic rays resulting from a solar storm (Figure 11.4). Unfortunately, this will require us to develop the capability to move dirt around, in large quantities, on another world. This won't be easy, lightweight, or cheap. But it might well be a requirement for a long-term settlement or colony.

Electromagnetic radiation shields have also been considered. It has been

Figure 11.4 Large equipment will be required to move the lunar regolith on top of the habitat to a depth of several meters, in order to stop the most energetic cosmic rays. (Image courtesy of NASA)

known for over a century that, when a charged particle like a solar wind atom traverses a magnetic field, it experiences a force changing its direction of motion. The stronger the field, the stronger the force and the more likely that the atom will have its trajectory altered to avoid whatever is protected by the magnetic field.

Funded by NASA's Innovative Advanced Concepts (NIAC) Program, a team of researchers from ASRC Aerospace Corporation have proposed shielding using electric fields suspended above a lunar base as seen in Figure 11.5. Their idea is to inflate large (5-m-diameter) conducing spheres and then mount them above the lunar base. They would then be charged to a potential of about 100 megavolts (!!) and serve as a deflector screen for the fast-moving, high-energy radiation coming in from the Sun. While this will only affect those cosmic rays coming in from above the base, the mass of the Moon will provide protection from those coming in from space on the other side of its surface. The end result would be a dramatic reduction in the overall crew radiation exposure.

One of the problems with this idea – and there are many – is the strength of the field required to bend fast-moving solar wind atoms around the ship. The

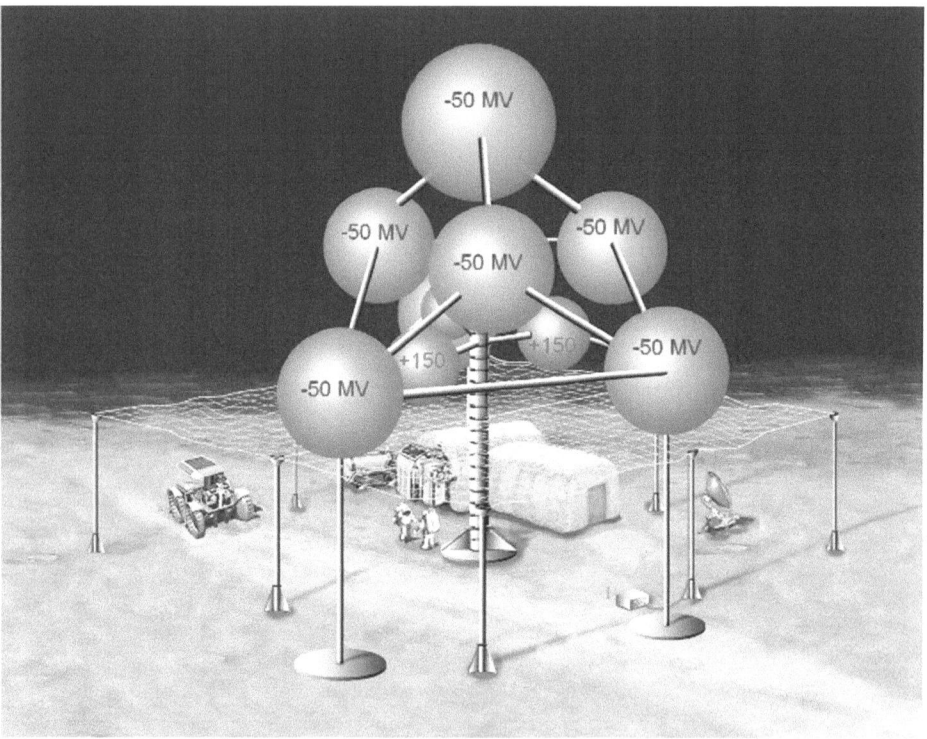

Figure 11.5 Artist concept of a lunar habitat protected by a high-voltage electrostatic shield. (Image courtesy of NASA and ASRC Aerospace)

power requirements are enormous and the weight of the wires and supporting power systems makes the overall approach not practical.

And we don't want to forget about the impact to the ISS resulting from a war in space. As you might imagine, the detonation of nuclear weapons in space would result in the damage or destruction of the ISS from the electromagnetic pulse (EMP) as described in Chapter 2, enhanced radiation exposure, or increased debris collisions resulting from the destruction of other satellites in similar orbits. War in space, as on the Earth, is certainly something to be avoided.

References

[1]   Wall, M. Space Station's Brush with Space Junk Highlights Growing Threat. Space.com, 29 June 2011, available online at *www.space.com/12107-space-junk-threat-growing-space-station.html*.

[2]   Space Debris and Human Spacecraft. NASA website, available online at *www.nasa.gov/mission_pages/station/news/orbital_debris.html*.

[3]   Near Miss: ISS Narrowly Escapes Debris Disaster. RT.com, 24 March 2012, available online at *http://rt.com/news/iss-wreckage-hazard-evacuation-359/*.

[4]   *NASA Orbital Debris Quarterly News*, **16**(2) (2012).

[5]   Bergin, C. Cupola Hit by Minor MMOD Strike, Shutter Closed for Evaluations. NASA Spaceflight.com, 12 June 2012, available online at *www.nasaspaceflight.com/2012/06/cupola-minor-mmod-strike-shutter-closed-evaluations/*.

[6]   Butikofer, R.; Fluckiger, E.; Desorgher, L.; Moser, M. The Extreme Solar Cosmic Ray Particle Event on 20 January, 2005 and Its Influence on the Radiation Dose Rate at Aircraft Altitude. *Science of the Total Environment*, **391**(2–3) (2008).

[7]   *NASA Science News* (27 January 2005).

[8]   Shea, M.; Smart, D.; McCracken, K.; Dreschhoff, G.; Spence, H. Solar Proton Events for 450 Years: The Carrington Event in Perspective. *Advances in Space Research*, **38**(2) (2006).

12 Effects on Scientific Research Satellites

If we were to suddenly lose all of our satellites, we probably wouldn't have the loss of space science data at the top of our list of immediate concerns. After all, unless you make your living designing, developing, or building satellites or are a scientist whose job it is to analyze the data coming back from a particular space science mission, then you may not even notice they were gone – at first.

But, as we on the Earth recovered from the economic disaster and human tragedy that would result from this catastrophe, their loss would slowly become more noticeable. We, as a people – and I don't just mean those in the so-called developed world, but all people – have shared in the journey of exploration wrought by the Space Age. If the disaster was caused by a runaway growth in the orbital debris population, then we might still be able to get information from those spacecraft not in Earth orbit – spacecraft like the venerable Voyagers or some of the relatively newly launched missions studying nearby asteroids and comets. If, however, the disaster was caused by a huge solar storm, then the loss might very well be Solar System-wide, potentially knocking out just about any spacecraft we have launched into space.

Earth-Orbiting Spacecraft (at Risk from Orbital Debris, Solar Radiation, and War)

Research satellites in Earth orbit are potentially vulnerable to all of the risks discussed so far: orbital debris, intense solar radiation, and war. Below are some prominent examples of satellite types we would lose if any of these were to occur.

Astronomy

When NASA decided to cancel the final planned servicing mission to the Hubble Space Telescope (HST) in 2004, the last thing the space agency expected was thousands of schoolchildren writing to NASA asking that the telescope be saved. NASA learned that the Hubble wasn't just their telescope – it is (verb tense change is intentional) the world's telescope and the public wasn't going to let it die a premature death. With heavy public pressure, NASA changed its policy in 2006 and the servicing flight was flown using the Space Shuttle orbiter in 2009.

The HST has been returning stunning images of the cosmos since the early 1990s. This 11,000-kg beast is the size of school bus and is in a 559-km orbit above the Earth. As of this writing, the HST researchers have produced nearly 10,000 papers describing the results of their research [1].

Yet, even with the best engineering and meticulous operation, the lifetime of the HST is limited. If NASA's current timeline holds, then the HST will be decommissioned some time before the year 2020. Regardless of an event causing its unexpected loss, the normal lifetime of spacecraft will eventually cause us to lose the HST. But, if the plans hold, we will have another telescope in place before that happens – a spacecraft that can pick up where HST leaves off and can continue its tradition of discovery.

NASA's Chandra X-Ray Observatory is considered to be a sister of the Hubble Space Telescope, observing the universe in light our eyes cannot directly see: X-rays. Chandra examines X-rays emitted from the very hot regions of the universe such as galaxy clusters, exploded stars, and matter around black holes. X-ray light won't penetrate the atmosphere, so this school-bus-sized telescope orbits in space in an elliptical Earth orbit that takes it nearly one-third of the way to the Moon. Just to give the reader an indication of the capability that would be lost if space telescopes such as Chandra were to suddenly be lost and/or unable to be replaced, it is interesting to note these facts about the telescope:

- Chandra can observe X-rays from clouds of gas so large that it takes five million years to go from one side to the other.
- Chandra's resolving power is equivalent to being able to read a stop sign from 12 miles away.
- The light from some of the objects observed by Chandra will have been traveling through space for ten billion years.
- Chandra can observe X-rays from particles up to the last second before they fall into a black hole.

Now imagine space-based astronomy ending. No more Hubble, no more Chandra (the sister telescope of the Hubble, observing the universe in X-ray light), and no new stunning pictures like those that have permeated our culture and that we take for granted.

Space Physics and Heliophysics
While not nearly as famous as Hubble, Earth-orbiting spacecraft like Hinode will also be threatened. Hinode, which means "sunrise" in Japanese, continuously observes the Sun, seeking to better understand it and, as a spin-off, help us to better predict its future activity. Hinode orbits the Earth in what is called a Sun-synchronous orbit – meaning that it doesn't often pass through the Earth's shadow and has a nearly continuous view of the Sun. Hinode is grouped with other Sun-observing satellites to support a field of study known as heliophysics [2]. ("Helio" comes from the Greek word for "sun".) It is somewhat ironic that the very object of its studies, the Sun, may one day destroy it and other Sun-observing satellites in a major solar storm.

There are also many satellites in orbit studying the physics of the ionosphere, the magnetosphere, and the near-space Earth environment in general. All would be lost. Humanity will have lost more than the tangible.

Earth Science
Science satellites don't just look outward; many, if not a majority, look down towards the Earth as we work to get a better understanding of the science right here at home. Examples include:

- Aqua: studying the Earth's water cycle, answering key questions about moisture in the atmosphere and our dependence upon the global water cycle;
- Aquarius: mapping the salinity of the oceans, the Aquarius will provide global information about humanity's impact on what is perhaps our most important resource;
- Aura: designed to study the Earth's atmospheric chemistry and its complex interactions, the Aura will provide information on global air quality, pollution levels, and changing atmospheric composition;
- Cloudsat: launched in 2006, the Cloudsat spacecraft has been studying the Earth's clouds – their formation, evolution, and relationship with climate and climate change;
- Global Precipitation Measurement (GPM): this collaborative project between NASA and Japan will provide next-generation observations of rain and snow worldwide every three hours. The data provided will be used to unify precipitation measurements made by an international network of partner satellites to quantify when, where, and how much it rains or snows around the world;
- GOES-N: technically a weather satellite, GOES-N also studies the atmospheric "triggers" for severe weather conditions such as tornadoes and hurricanes;
- Ice, Cloud, and Land Elevation Satellite: the satellites in this ongoing program will measure the mass of the Earth's ice sheets, the height of clouds and suspended atmospheric aerosols, land topography, and vegetation;
- Tropical Rainfall Measuring Mission (TRMM): a satellite designed to aid in our understanding of the water cycle and the role it plays in climate; by covering the tropical and semi-tropical regions of the Earth, TRMM provides much-needed data on rainfall and the heat release associated with it.

Questions about our world and the impact of humans upon it will remain indefinitely unanswered if we were to lose access to Earth orbit.

Deep-Space Spacecraft (at Risk from Solar Radiation)

Our research spacecraft working beyond Earth orbit are not vulnerable to orbital debris and it is unlikely that they would be targeted during a war. That leaves them only vulnerable to unexpectedly high solar radiation. Their loss, therefore, is fortunately not as likely as that of our other spacecraft.

The first spacecraft likely to go silent in this scenario would be those in the inner Solar System, like MESSENGER (Mercury Surface, Space Environment, Geochemistry and Ranging). Launched in 2004, MESSENGER took a rather roundabout trip to Mercury, having to swing by Venus twice, Mercury three times, and even back around the Earth once to place it in Mercury's orbit. It is noteworthy to mention that MESSENGER is the first spacecraft ever to orbit Mercury. All other spacecraft sent to this tiny planet so close to the Sun have merely flown by. MESSENGER has sent back to the Earth over 100,000 images of the planet (Figure 12.1) and data that will help scientists to understand it and its formation [3].

Next to be lost would be the Advanced Composition Explorer (ACE). ACE, which provides the world's early warning for solar storms, orbits the Sun in a

Figure 12.1 MESSENGER captured this never-before-seen area of Mercury's South Pole. (Image courtesy of NASA/Johns Hopkins University Applied Physics Laboratory/ Carnegie Institution of Washington)

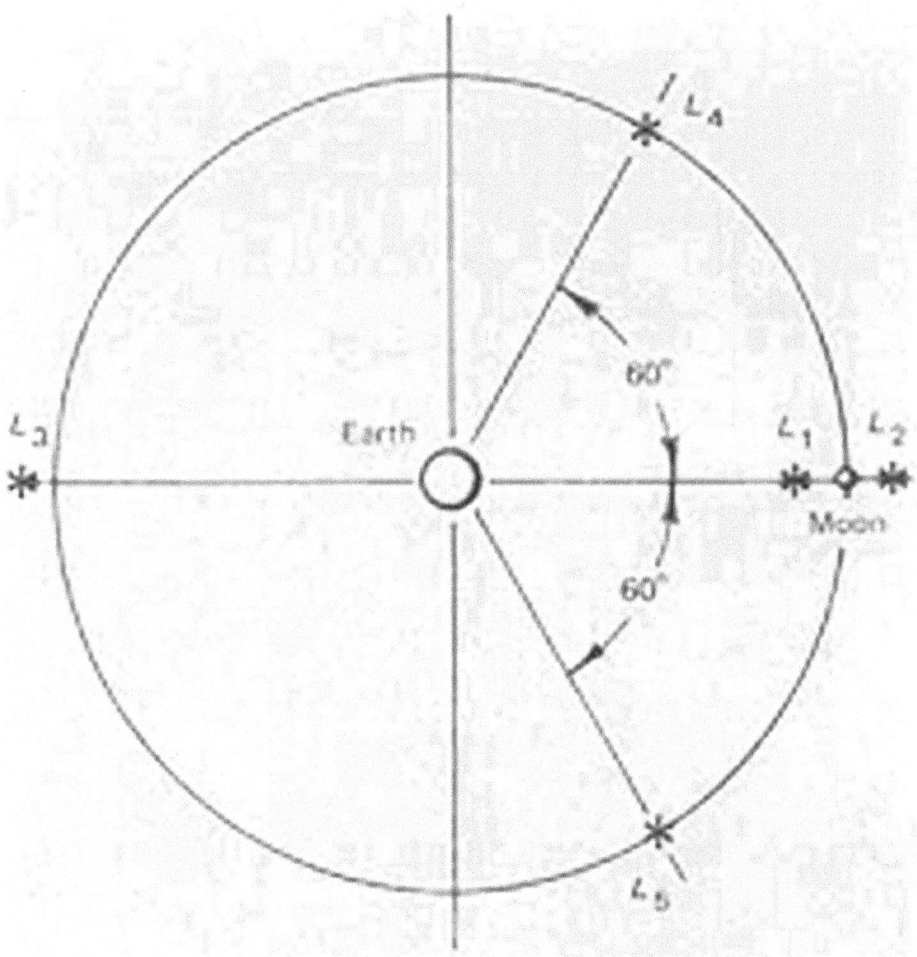

Figure 12.2 Libration points are regions of space where the gravitation pull of the Earth and the Sun make it relatively easy to place a spacecraft so that it will co-orbit the Sun with the Earth and require little fuel to maintain its position. (Image courtesy of NASA)

region of gravitational equilibrium known as the Earth/Sun L1 libration point (Figure 12.2). In the Solar System, L1 is prime real estate if you want to observe the Sun and be in a direct line with the Earth at the same time. Located about 1.5 million km from the Earth and just under 150 million km from the Sun, ACE studies the stream of radiation coming from the Sun known as the solar wind and can detect increases in solar radiation output, such as those found in coronal mass ejections and solar flares.

Currently, when ACE detects an increase in the radiation coming from the Sun, it sends a warning back to the Earth so that satellite operators can prepare

Figure 12.3 The NASA Opportunity Rover found its own crashed heat shield as it traversed the Martian surface. (Image courtesy of NASA)

their spacecraft for the increased radiation levels before they actually arrive. The warning is sent by radio, traveling at the speed of light, which is much faster than the particles of radiation that triggered the alert. Warning times are typically about an hour.

Mars is close enough to the Sun that any storm intense enough to damage spacecraft here at the Earth would likely also threaten those orbiting there. As of this writing, there are three operational spacecraft orbiting Mars and rovers on its surface (Figure 12.3). The satellites orbiting Mars are studying the planet, its composition, its surface, and its atmosphere in unprecedented detail. We're very close to finding out whether Mars once had conditions suitable for the development of life and whether it might be able to support Earth-like life today. Mars has often been referred to as a "sister planet" of the Earth. Careful and long-term study of our sister planet will certainly help us to better understand our own planet, Earth.

A personal favorite of mine, thanks to the innovating electric propulsion system that propels it through space, leaving a small trail of ionized gas, the Dawn spacecraft will fly to, and rendezvous with, the asteroid Vesta and dwarf planet Ceres. Both of these main belt asteroids are believed to have accreted early in the history of the Solar System. The mission will characterize the early Solar System and the processes that dominated its formation.

Figure 12.4 This image was captured on Cassini's approach to Saturn's icy moon, Enceladus, flying a mere 41,800 km above its surface nearly 1.2 billion km from the Earth. (Image courtesy of NASA)

Moving farther out into the Solar System, we find that the next spacecraft to possibly be lost would be the Juno, on its way to Jupiter. The primary goal of the mission is to help us understand how Jupiter came to be the giant planet that it is. Hidden beneath its massive cloud layers, the planet may contain answers to some of our fundamental questions about the origins of the Solar System. In addition, with the discovery of many extrasolar planets similar to it, Jupiter may hold the key for understanding the formation of these other solar systems as well.

Launched way back in 1997, the Cassini spacecraft is alive and well, still performing cutting-edge science as it circles Saturn and regularly swings by its many moons and elaborate rings. The spacecraft is performing very well and its mission will be extended until at least 2017, if everything continues to function as it has. Without Cassini, and missions like it, we would forever know Saturn's moons as dim specks of light in a telescope rather than the exciting and interesting worlds that they are (Figure 12.4).

Launched when Pluto was still considered to be a planet, the New Horizons mission will fly by this denizen of the outer Solar System and provide us with detailed understanding of it – and perhaps give us insight into the many, many worlds like it that populate the outermost regions of the Solar System.

Fortunately, these outer Solar System probes are not as likely to be imperiled as closer to the Sun. The intensity of solar storms will certainly decrease with distance and, if there is one thing the Solar System has in abundance, it's empty space!

References

[1]   Hubble Space Telescope Science Institute press release. Hubble Racks Up 10,000 Science Papers (6 December 2011).
[2]   Kosugi, T., et al. The Hinode (Solar-B) Mission: An Overview. *Solar Physics*, **243**(1) (2007).
[3]   Solomon, S.C., et al. The MESSENGER Mission to Mercury: Scientific Objectives and Implementation. *Planetary and Space Science*, **49**(14–15) (2001).

Part 3

What Can We Do?

13 Reduce the Growth in Orbital Debris

According to the aerospace consulting group Euroconsult, an average of 122 satellites will be launched into space every year for the next decade [1]. In their report, "Satellites to Be Built and Launched by 2019, World Market Survey", they estimate that these satellites represent over $194 billion in revenue. Each and every one of these satellites is a potential orbital debris disaster waiting to happen. Fortunately, many of these newly launched spacecraft will abide by international standards developed to limit the growth in orbital debris.

It is NASA policy for all of its satellites to comply with something called NASA Safety Standard 1740.14. (Yes, there are probably at least that many NASA policies. Fortunately, many of them are probably old and no longer enforced.) The standard says that it is NASA's policy to "employ design and operations practices that limit the generation of orbital debris, consistent with mission requirements and cost-effectiveness". The first clause sounds pretty good: NASA will design its future missions and spacecraft to limit the growth of orbital debris. The second clause gives project managers some wriggle room and possible ways to not comply with the standard: consistent with mission requirements and cost. In other words, NASA will engineer its spacecraft to assure that they don't increase the orbital debris problem unless, of course, it is going to keep the mission from succeeding or is too expensive.

NASA's first guidelines for limiting the growth of debris were adopted in 1995 and, a few years later, the rest of the US government adopted similar guidelines. Now, all US government satellites, including those launched by the Department of Defense and the National Oceanic and Atmospheric Administration (NOAA), have to be designed to limit the growth of orbital debris. Other countries and agencies, including the European Space Agency (ESA), Japan, and Russia, now have similar guidelines. Finally, in 2007, the Scientific and Technical Sub-committee of the United Nations Committee on the Peaceful Uses of Outer Space reached consensus on an international set of guidelines similar to NASA's [2]. The United Nations adopted the guidelines in 2008.

So, now we have the major space agencies and governments of the world in agreement that orbital debris is a problem. What do these new guidelines mean in practical terms? How do you design a spacecraft to limit the growth of orbital debris?

Deplete On-Board Energy Sources

Translation: Don't let things explode
This requirement doesn't just apply to a spacecraft. In fact, many of the explosions that generated debris came from the failure of pressurized tanks on the rockets carrying the satellites into space. Rockets burn fuel to produce thrust and that fuel is, by its very nature, volatile and often explosive. To safely contain the fuel before it is used by the rocket, it is stored in pressurized tanks. The tanks used to be designed to last just long enough to get the rocket and its payload into space and then, well, then the mission was over and no one really cared whether or not the tank failed. After all, the tanks would be empty of fuel, right?

Wrong. Placing a satellite safely in its proper orbit is not the exact engineering feat many think it is. There are many variables that affect how much fuel will be required to put a satellite where it needs to be, including how much fuel is needed to simply reach space (sometimes it takes a little more fuel, sometimes less), how accurately the rocket's guidance system can place the satellite in the correct orbit (additional fuel may be needed to correct for initial guidance mistakes), and how much of a fuel margin the launch provider wants to carry "just in case" it is needed. Launching satellites is a big and expensive business. Would you want to be the company that puts a $1 billion spacecraft into the wrong orbit just because you decided to not carry enough extra fuel to correct for some simple targeting mistake?

The outcome of taking into account these variables is the loading of additional fuel onto the rocket that may or may not be needed to successfully complete the mission. Most of the time, there is fuel left over in the tanks. Herein lies the problem. After a few years in space, even the best-designed tanks may begin to leak or simply fail catastrophically, causing an explosion of the rocket stage and generating additional space debris.

The new international standards require rocket providers to have a way to safely vent unused fuel, relieving the pressure in the tanks to lessen the possibility that they will explode.

What else might explode?

Rockets often launch and fly over populated areas. Although most countries try to limit this type of thing by launching out and over the open ocean, some launch sites are land-locked and flying over inhabited land is unavoidable. Sometimes, rockets go off course or fail early in flight, long before they reach space. To avoid the disaster that would occur should a malfunctioning rocket fall on a city or other populated area, they carry "range safety" explosives to allow them to be exploded early in flight and over areas where the damage will be minimal. Who can forget seeing those famous images of rockets exploding just after they cleared the launch pad? Many of those explosions were not accidental; range safety officers who saw that the rocket was malfunctioning and a risk to people downrange triggered them. In such cases, the range safety officer triggered the on-board explosives to destroy the rocket and limit the damage.

If the rocket doesn't fail and successfully reaches space, proper precautions need to be made to ensure that the unused safety explosives don't accidentally go off and destroy the rocket while it is in orbit.

Batteries can also explode. A few years ago, the media were full of stories about computer laptop batteries overheating or exploding [3]. We should not be surprised. Almost all batteries store energy chemically. The better the battery, meaning the longer it lasts without recharging or being replaced, the better it is at storing energy. This also means that the best batteries are those that store the most energy. The difference between a battery and a bomb is, to be simplistic, the rate at which they release energy. A bomb destructively releases most or all of its energy in a very short amount of time. A battery can contain the same amount of energy but, if it releases it in a slower and more controlled way, then we use it in our laptops, cameras, and, yes, on our spacecraft.

To reduce the risk of batteries exploding, aerospace companies and space agencies require lots of analysis and testing of batteries. Sample batteries are built and then tested under almost any conceivable condition they may encounter in their lifetime – from manufacture, to shipping, to long-term storage in a warehouse, to launch on the rocket and flight in space. Batteries are heated to extreme temperatures to see when they explode. They're dropped to see whether they will fail on impact. They're discharged rapidly to see how they behave under different failure scenarios. And the testing doesn't stop when a good design is found. Random tests of batteries in storage are conducted to make sure no manufacturing defects occurred in any particular set waiting to see use in space.

Another item that may fail and cause an explosion is something called a control moment gyroscope (CMG). A CMG is basically a spinning wheel that is used to help a spacecraft alter its orientation once it is in space. This is possible due to the physical property known as conservation of angular momentum. Basically, a spinning object likes to retain its orientation better than a non-spinning one. The best example is a spinning bicycle wheel. Have you ever noticed that it is easier to fall over on a bicycle before you begin pedaling than after? A moving bicycle wheel wants to keep its orientation, with its spin axis parallel to the ground, allowing you to remain on it and not fall over. When at rest, the wheel is not spinning and no force is acting to keep the bicycle from falling over.

CMGs on spacecraft can rotate very rapidly and, if something were to go wrong with one, then it can fly apart, damaging the surrounding spacecraft.

Limiting Post-Mission Orbital Lifetime to 25 Years

Translation: Throw away your trash when you are finished using it
There are thousands of long-dead satellites orbiting the Earth that could easily have been removed from orbit had their designers planned for them to be capable of doing so. Most didn't – and we are living with the consequences. Within 25 years of the end of their useful mission life, new spacecraft must be

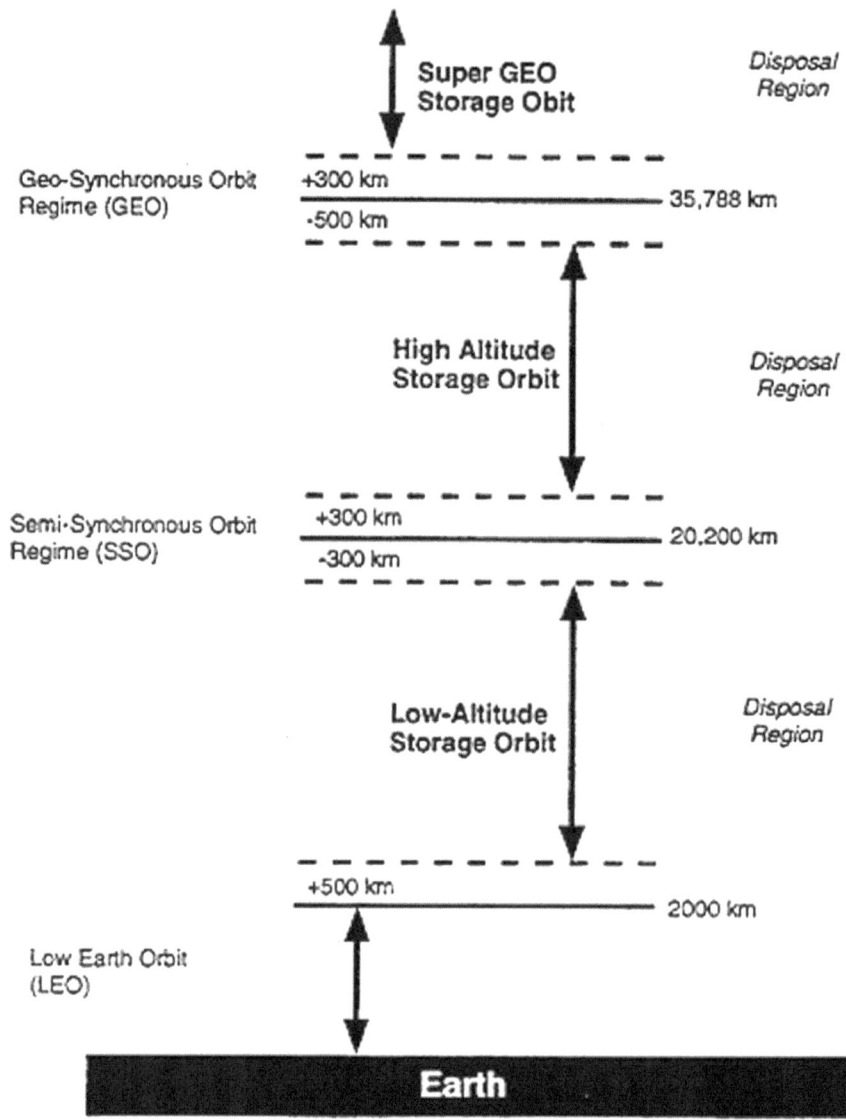

Figure 13.1 Earth-orbit regions and the possible disposal locations that meet NASA's orbital debris-mitigation requirements. (Image courtesy of NASA)

designed to either return from space, de-orbit, or to go to an orbit in which they are not a threat (Figure 13.1).

For satellites that end their mission life in low-Earth orbit (LEO), they must:

1. place themselves in an orbit that will naturally decay until the spacecraft re-enters the more dense atmosphere and burns up; satellites orbiting in LEO

will have their orbits decay due to the small but non-zero tenuous atmosphere there producing a drag force on the spacecraft, causing it to lose energy and slowly spiral downward to lower and lower orbits;

2. use on-board propulsion to boost to a so-called storage orbit between LEO and geostationary Earth orbit (GEO); specifically, this means the satellite must be placed into an orbit above 2,500 km and below 25,000 km at the end of its useful life; satellite manufacturers choosing this option must take along extra propellant and design their propulsion systems to be operable after several years on orbit;

3. be retrieved somehow; NASA could do this during the Space Shuttle era by using the robotic arm to bring old spacecraft into the Shuttle's cargo bay

Figure 13.2 NASA's Space Shuttle deployed and retrieved the Long Duration Exposure Facility (LDEF) satellite from space. (Image courtesy of NASA)

and return it to the Earth; this capability was demonstrated by the retrieval of NASA's Long Duration Exposure Facility (LDEF) by the Space Shuttle *Columbia* in 1990 (Figure 13.2); the LDEF was launched into space by the Shuttle *Challenger* in 1984 and was designed to be deployed and retrieved by the Space Shuttle; most satellites are not designed to be grappled and brought home by another spacecraft and, unfortunately, there are no spacecraft flying today that could capture and bring them home even if they were.

For missions that end their lives somewhere above LEO (above 2,000 km), they must boost either to a storage orbit above GEO or into the one located above LEO and below GEO.

Oh yes, the satellite manufacturer has to show that these requirements can be met with a greater than 99% probability.

Limiting the Creation of New Debris

Translation: Don't make a mess
In the early days of space exploration, a satellite that needed to have an instrument covered during launch to protect its sensitive optics or electronics would simply jettison the lens cover into space when it was time to begin using the instrument. That hypothetical lens cover is probably still orbiting the Earth at 7 km/sec. Under the new guidelines, that same lens cover would likely now be hinged, allowing the instrument to be protected during launch and opened for viewing when in space – without generating any new debris.

Rocket upper stages, like the one used to loft the world's first satellite, Sputnik-1, used to become part of the debris population as a matter without question. The 30-m-long and 4,000-kg stage that placed Sputnik-1 into orbit remained there with it, becoming the world's first piece of orbital debris. Today, rocket launch providers are required to limit the orbital lifetime of these debris objects just as they limit the orbital lifetime of operational satellites after their mission is completed, as described above.

Still, accidents happen. In 1965, astronaut Ed White took America's first extravehicular activity (EVA) and, amazingly enough, accidentally let go of an extra glove. The glove promptly went into its own orbit around the Earth [4].

Not all debris is accidental. Some of *Star Trek* creator Gene Roddenberry's ashes were carried into space and left there. For some reason, probably because I am such a dedicated *Star Trek* fan, I contend that this is one piece of debris that deserves to be there! Although, in all fairness, the ashes and their container didn't remain in orbit very long – they entered the Earth's atmosphere and burned up within just a few years of launch [5].

Limiting the Consequences of Impact with Orbital Debris

Translation: Design your spacecraft to be more robust
New spacecraft can be built to withstand, or at least partially withstand, a collision with small debris objects in space. To do this effectively, a thorough analysis is conducted, taking into account some of the following considerations:

- What is the probability of being hit by a debris object?
- If the satellite is hit, what is the probably that the debris object will penetrate as a result of the impact?
- What is the probability that the penetrating impact will disable or destroy the satellite?

This kind of analysis is required for all crewed spacecraft. According to a National Research Council (NRC) report on orbital debris and micrometeoroid protection, the placement of the Space Shuttle payload bay doors was determined to mitigate the risk of an impact with some pressurized tanks stored in the payload bay during the STS-73 mission [6]. According to the NRC, the doors sustained a non-penetrating impact that might have been catastrophic had the doors remained open and the debris object impacted one of these pressurized tanks.

Various design options exist for shielding that can survive collision with small pieces of debris, allowing the spacecraft to continue functioning even after a collision. These micrometeoroid and debris shields can be tested for their effectiveness on the ground using specialized guns that accelerate small objects to speeds approaching orbital velocity to determine the designs that are most effective and affordable. Selecting a debris shield is a balance of effectiveness against the debris objects most likely to be of concern for a particular mission, mass (a shield weighing as much as the spacecraft itself is not likely to be a viable option), and cost.

Unfortunately, we are limited in what we can do to mitigate the effects of a collision with orbital debris. If a satellite collides with another spacecraft or spent rocket stage, like the Iridium Cosmos collision described in Chapter 1, there really isn't much that can be done – even if the spacecraft or satellite were designed to withstand impacts with small debris objects. Nothing this author can imagine would be capable of sustaining and surviving an impact with a large research or telecommunications satellite.

Limiting the Risk on the Ground

Translation: Don't let your messy design kill someone
The often-preferred approach to limiting the growth of orbital debris is to force the spacecraft or rocket stage to de-orbit and fall back to the Earth. This sounds good until you consider that parts of the satellite might well survive their passage through the atmosphere and impact something on the ground – hopefully not

you or your property! While being hit by a piece of space junk is not likely to happen, it is not impossible.

In 1978, a near-nightmare scenario nearly played out in Canada when Cosmos-954, a Soviet satellite with an on-board nuclear fission reactor, fell to the Earth. Fortunately, the area where the debris fell, including radioactive debris, was sparsely populated and the debris field did minimal harm to the environment. Had it fallen in a major city, the outcome could have been very different [7].

More recently, NASA's Upper Atmosphere Research Satellite (UARS), launched in 1991, fell into the Pacific Ocean in September 2011. UARS, weighing 5,900 kg, likely had several of its components survive the passage through the atmosphere before impacting the ocean. Although thankfully not radioactive, the UARS de-orbit could have been calamitous had it impacted on land.

Space debris has been found on land and one piece is purported to have passed closely by an Oklahoma woman in 1997. According to reports, she felt something touch her shoulder before it hit the ground near where she was walking. The object was later found to be part of a rocket launched a year earlier [8].

Fortunately, the above examples are the exceptions and not the rule. Spacecraft are now designed to mostly burn up as they pass through the atmosphere, reducing to virtually zero the possibility of any pieces contained therein reaching the ground. Spacecraft parts are selected to accomplish their mission and to not survive re-entry. Those satellites too large to burn up or with significant piece parts too massive or large to burn up are placed on re-entry trajectories that will crash them into the ocean far away from civilization.

References

[1]   Euroconsult press release. More than 1200 Satellites to Be Launched over the Next 10 Years (6 September 2010).

[2]   NASA Orbital Debris Program Office website, *www.orbitaldebris.jsc.nasa.gov*.

[3]   HP Notebook PC Battery Pack Replacement Program, available online at *bpr.hpordercenter.com/hbpr/*.

[4]   Crowther, R. Space Junk: Protecting Space for Future Generations. *Science*, **296**(5571) (2002).

[5]   Trex, E. 10 Bizarre Places for Cremation Ashes. CNN.com, 2 August 2010, available online at *www.cnn.com/2010/LIVING/08/02/mf.cremation.ashes.-where.go/index.html*.

[6]   Committee on Space Shuttle Meteoroid/Debris Risk Management, Aeronautics and Space Engineering Board, Commission on Engineering and Technical Systems, National Research Council. *Protecting the Space Shuttle from Meteoroids and Orbital Debris*. National Academy Press, Washington, DC (1997).

[7]   Heaps, L. *Operation Morning Light: Terror in Our Skies: The True Story of Cosmos 954*. Paddington Press, New York, NY (1978).

[8]   NASA-HANDBOOK 8719.14. Handbook for Limiting Orbital Debris, 2008-07-30.

14 Reduce the Amount of Debris in Space

We've put over half a million pieces of junk in Earth orbit that imperil our use of space now and in the future, so what can we do about it? The answer, surprisingly, is that we can do a great deal should we decide it is worth the effort.

Large Debris Removal

Studies by NASA, the European Space Agency (ESA), and the United Nations indicate that the removal of five large debris objects per year might be enough to slow down the growth of orbital debris and, over time, dramatically reduce the probability of on-orbit collisions [1]. The thinking is that, by removing the big pieces of junk, you can reduce the probability of these large objects colliding and producing thousands of smaller debris pieces. Looking at the population of debris objects, a large number of them are old Russian rocket stages with masses of up to 8.3 tons and they are anywhere from 6 to 12 m in length [2]. Many others are old, dead spacecraft long past their mission lives.

To successfully remove these large objects, a space system has to perform the following minimum functions:

1. Launch and maneuver in close proximity to the debris object to be removed.
2. Rendezvous with the target and fly in close formation with it.
3. Stabilize the target for capture.
4. Capture the target and attach to it the de-orbit or orbital maneuvering system.
5. Optional: detach and move to the next orbital debris target.

Some approaches described below have their own, unique, requirements that will be noted as appropriate.

Conventional Orbital Tugs

The least-risky approach is to use as much existing technology as possible in a spacecraft that will rendezvous and de-orbit the large debris object of interest. This probably means using a chemical or electric propulsion system to perform the in-space maneuvers required, close-proximity operations, and the de-orbit itself. The de-orbit could be as simple as the tug grabbing the object and then firing its thrusters to send it and its now-captive payload into a controlled de-orbit for impact somewhere at sea. It might also carry the object to one of the

Figure 14.1 The NASA/DARPA Orbital Express mission demonstrated the automated rendezvous and capture of a cooperative target in 2007. (Image courtesy of NASA and DARPA)

higher disposal orbits discussed in Chapter 13. In either case, it will require lots of fuel.

The tug would perhaps take advantage of the automated rendezvous and capture technologies demonstrated by NASA and the Defense Advanced Research Projects Agency (DARPA) in the Orbital Express mission flown in 2007 (Figure 14.1).

Orbital Express had as its goal the demonstration of on-orbit autonomous satellite servicing. The project demonstrated many technologies that will be required for orbital debris mitigation, including rendezvous and close-by operations, satellite capture, and manipulation of the target satellite by the servicing satellite. A key component of the successful demonstration was the grappling and capture of the target satellite. Orbital Express proved that this sort of mission can be accomplished, the only caveat being that the target satellite was "cooperative", meaning that it was stabilized and designed to be captured. It remains to be seen how such a system might catch a tumbling satellite not otherwise designed for on-orbit servicing. But it was a successful first step [3].

If the grappling problem can be solved (see below), then this is clearly a relatively low-risk approach to removing debris. There is a significant problem, however. Conventionally propelled (chemical or electric) space tugs cannot carry enough fuel to accomplish a rendezvous-and-capture maneuver with more than one or two objects before running out of fuel. Chemical propulsion will likely enable a spacecraft to de-orbit one piece of large debris. Studies indicate that a tug propelled by solar electric thrusters might be able to remove two or three objects before it, too, runs out of gas. With each spacecraft costing between $300 million and $500 million, this is clearly shaping up to be an expensive garbage-removal service. Can we do anything that might allow a spacecraft to de-orbit many objects during its lifetime, driving the cost of removing each object down? If you are willing to consider some new, moderate-risk technologies, then the answer is "yes".

Electrodynamic Tether-Propelled Tugs
Nature has provided us with a way to take advantage of the natural space environment for propulsion, providing an opportunity to develop spacecraft that can operate in low-Earth orbit (LEO), maneuvering between orbits at will, consuming no propellant: electrodynamic tethers (EDTs) (Figures 14.2 and 14.3). EDTs are not rockets and are therefore not limited by the amount of fuel they can carry. Rather, they use an electrodynamic interaction between a current-carrying wire and a magnetic field to produce thrust. The reason an EDT is interesting is because electromagnetism provides the propulsive force rather than the expulsion of propellant.

Figure 14.2 Electrodynamic tethers provide propulsion without fuel, enabling the de-orbit of multiple debris objects by a single spacecraft. (Image courtesy of NASA)

A long conducting wire, or tether, can be used as an EDT thruster. By long, I mean long – several kilometers in length. A wire carrying an electric current experiences a force when it is in the presence of a magnetic field. The force acts in a direction perpendicular to both the direction of current flow (which is constrained to be in the wire) and the direction of the magnetic field. This type of propulsion is just a clever way of getting an electric current to flow in a tether so that the Earth's very strong magnetic field will accelerate the wire and anything attached to it – such as our spacecraft [4].

EDTs will only work in low orbits, below about 2,000 km, due to the drop-off in the density of the ionosphere. The ionosphere is essentially a cloud of ionized atmosphere, meaning it is atmospheric atoms that have been stripped of electrons. These stripped electrons, mixed together in roughly equal proportions with the atoms from which they came, form what is known as plasma. Combining the ionosphere with the Earth's magnetic field is the "perfect combination" of environments to allow EDTs to function. Figure 14.4 shows the elements of an EDT thruster and its operation in the LEO environment.

An EDT-propelled spacecraft should be capable of flying to a debris object, grappling and capturing it, attaching or accomplishing its disposal, and then flying on to the next object for a repeat performance. In theory, such a system could be used over time to dispose of tens or hundreds of debris objects, never running out of fuel [5].

This approach is not without drawbacks, however. For one, the tether, no matter how well controlled, will pose a collision risk not only with the target satellite, but with other objects in space during its mission lifetime. Recall that the tether is kilometers in length, making it among the largest objects ever flown in space by far. Next, the tether will be librating, or gently swinging (within only a few degrees – we're not talking about 90° swings here!), throughout its operation. Can a spacecraft with a multi-kilometer tether swinging above it successfully rendezvous and capture another spacecraft? That remains to be proven.

EDTs can also be combined with chemically or electrically propelled spacecraft to accomplish the de-orbit maneuver, sparing the tug from having to expend some of its fuel and potentially extending its on-orbit lifetime. In this scenario, a small, self-contained EDT thrust package is attached to the debris object during the rendezvous. After the tug moves away, the EDT deploys and begins thrusting, lowering the attitude of the debris object until it enters the Earth's atmosphere and burns up. These small EDT thrust packages have been studied by several companies and appear to be feasibly very lightweight and inexpensive to build [6].

Enhanced Drag Devices

Low-Earth orbit is not a perfect vacuum. It contains the ionosphere (described above), which, in addition to providing electrons for operating EDT thrusters, causes a drag force on objects passing through it in the same way as the atmosphere near the Earth's surface causes drag on an automobile. This drag

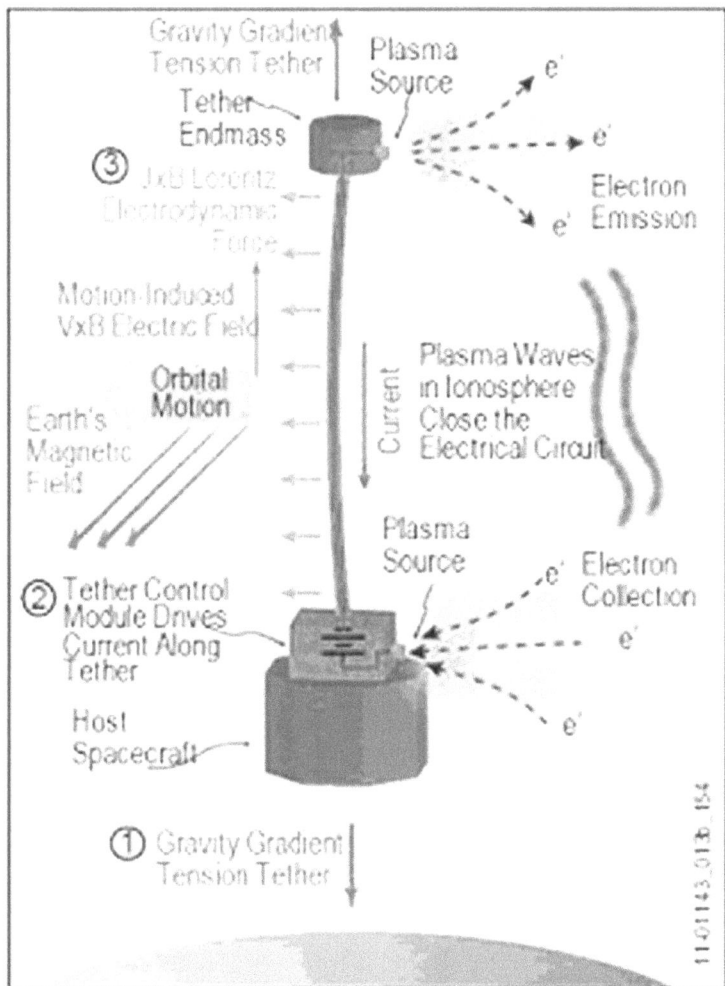

Figure 14.3 An electrodynamic tether (EDT) propulsion system derives its thrust from interaction with the Earth's magnetic field and ionosphere. Electrons captured from the Earth's ionosphere flow through the tether wire and are emitted back into the plasma from the other end. Electrons have a charge, so they experience a force when they move through the Earth's magnetic field. Since they are constrained to flow through the tether wire, the force is transferred to the wire and the attached spacecraft, producing thrust. (Image courtesy of Tethers Unlimited, Inc.)

force can be significant, causing the orbits of objects a few hundred kilometers above the Earth to naturally decay until the object enters the atmosphere. The rate at which an object is affected by this drag depends upon many factors, including its overall mass and surface area. The heavier the object, the longer it will take the drag forces to slow it down sufficiently for atmosphere entry. The larger its surface area, the more it will be slowed by interaction with the residual

Figure 14.4 NASA's NanoSail-D showed that a small deployable sail can increase the aerodynamic drag on satellites deployed in LEO. (Image courtesy of NASA)

atmosphere in space. The orbit of a large, lightweight object will generally decay faster than that of a smaller, heavier object. If it were possible to take an existing small, relatively heavy satellite and increase its surface area, then its orbit would decay faster, naturally removing it from space and as a debris object.

One way to accomplish this would be for an orbital tug to grapple and attach something to the debris object that would subsequently increase its atmospheric drag. There are several ideas for what this enhanced drag devise might be, including a drag sail and, simply enough, a balloon.

A drag sail is basically a conventional sail like you might see on a sailing ship crossing a lake. The premise is simple. The sail folds into a small box that is attached to the debris object by the orbital tug. Once the sail is deployed, the combined drag of the sail and the debris object will increase, causing the object's orbit to decay and eventual re-entry. The sail, folded to fit within a very small area before deployment, might attain a surface area of several tens or hundreds of square meters, dramatically increasing the aerodynamic drag of whatever it is attached to.

NASA demonstrated the deployment of such a device with the Earth orbital flight of the NanoSail-D in 2010 (Figures 14.4 and 14.5). The NanoSail-D was deployed from a very small satellite weighing only a few kilograms and unfurled to be 10 square meters in area [7]. One can envision our orbital tug carrying several stowed NanoSail-D devices in a rack, with the robotic arm attaching one from the stack to each debris object slated for removal.

Figure 14.5 NASA's NanoSail-D "drag sail" in a ground-deployment test prior to flight. The sail's deployed area was about 10 square meters and it was autonomously deployed from the small CubeSat shown in the center of the sail. (Image courtesy of NASA)

Again, there are drawbacks. For one, the sail needs an unobstructed area in which to deploy. If the target is a spacecraft with multiple antennas, rocket engines, or instrument booms, then there is a risk of the sail being shredded during deployment by snagging on one of these deployed structures. In general, however, it is an elegant and simple solution that has already been demonstrated in space.

Balloons have also been demonstrated in space and offer yet another relatively simple method of increasing an object's aerodynamic drag, forcing an earlier-than-otherwise-possible de-orbit and disposal. The scenario is similar: the tug grapples with the debris object and places upon it an un-inflated balloon package, including the tank containing the pressurized gas that will inflate the balloon. This basic technology was demonstrated in space very early in the space program in the Echo balloon series of missions that began in 1960 (Figure 14.6) [8]. Balloons measuring 135 feet (41 m) in diameter were flown into space, each requiring only a few pounds of gas to inflate to this enormous size. For comparison, inflating the balloon on the ground required 40,000 pounds of air! Much smaller and lighter-weight balloons can be built today for de-orbit applications.

The risks associated with using inflatables for de-orbit applications are similar to those with drag sails – interference by booms, arrays, and instruments that

Figure 14.6 A large, inflatable balloon similar to, although smaller than, the NASA Echo balloons launched in the early 1960s might be used to de-orbit large orbital debris objects. (Image courtesy of NASA)

might cause the balloon to shred or leak during deployment and inflation is chief among them.

Object Capture
The general problem facing all of the techniques and technologies mentioned above is that of capture. The debris object to be removed will be uncooperative: spinning, nutating, or tumbling orientations are all possible. Furthermore, the debris will likely have sharp protuberances such as antennas, thrusters, and solar arrays inconveniently placed around their exteriors, making capture using a robotic arm very risky. Clearly, another approach is needed and several have been proposed.

Imagine a spacecraft flying in close proximity to its target and, instead of reaching out and grappling it with a robotic arm, a net is cast, Spiderman-style, ensnaring the errant object and capturing it so a de-orbit propulsion module can move it out of the way [9]. This could be challenging with a rotating spacecraft, since the rotation itself would not slow after the net was cast; the space tug would then have the challenge of stabilizing the debris without losing control of its own attitude. There is also the risk that the net will break off one or more spacecraft parts, creating a cloud of smaller debris objects that must subsequently be dealt with.

Another approach for capturing the debris would be by using something to make either the boom or the debris sticky and therefore more easily grappled by a robotic arm. Since conventional sticky substances on the Earth don't typically work the same way in space, innovative engineers at Altius Space Machines are developing a robotic arm that uses a pad at its tip, instead of more traditional clamps, to induce electrostatic charges in the target body, allowing the arm to stick to the debris in the resulting electroadhesion. The process should work with just about any space material, independently of whether or not the material is conductive or insulating [10].

Small Debris Removal

The above techniques may work for the large debris, but what about the nearly half a million small debris objects that surround the Earth? (For this discussion, "small debris" refers to objects less than 10 cm in diameter.) We simply cannot affordably get rid of that many objects with a conventionally propelled spacecraft in any reasonable amount of time. They are too small to grab and de-orbit with a conventional spacecraft, regardless of how it is propelled and how good its grapple may be.

Perhaps the most technically promising solution to the small debris problem may also be the most controversial: lasers.

Ground and Space-Based Lasers
Imagine a video game in which you have to shoot down incoming missiles with your handy-dandy space laser or beamed energy weapon. On the first screen, the missiles come at you at a leisurely pace and are fairly easy to destroy. On the subsequent screens, they come faster and faster until you are simply over-whelmed with their speed and numbers, and then you die. Shooting down small debris, which is really a misnomer, since what will actually occur is a "bumping" of the debris, will be more like the second or third screens in the video game rather than the first or last. In other words, the debris objects will come quickly, but not so many at the same time as to be overwhelming.

A laser, which is really nothing more than a beam of intense light, can propagate through the vacuum of space very well. In the atmosphere, the laser light will scatter from atmospheric atoms, losing energy and spreading out with ever-increasing distance. In space, this is not so much of a problem. There is some residual atmosphere in LEO, but not enough to make it impossible for a laser beam to strike a piece of debris.

When people think about a high-energy laser striking a debris object, they imagine the beam obliterating the object, turning it into vapor and forever removing it from being a threat to another satellite. Unfortunately, it's not that simple.

The first thing to realize is that the small debris objects are not coming from any single direction. As discussed in Chapter 1, small debris is circling the planet at various altitudes and inclinations, and with varying speeds. We currently aren't able to track all of the small objects very well so their actual location and orbital elements are not precisely known. There are clusters of small debris in various altitude bands and, based on radar measurements and modeling, we only approximately know where these clusters are located.

The next thing to know about small debris is that there are many kinds out there, including, but not limited to:

- metallic nuts and bolts;
- shrapnel from exploded rocket parts, also typically metallic;
- foam insulation;
- frozen liquids of various types, from water to rocket propellant and human urine;
- radioactive molten salt droplets leaking from old Soviet-era space nuclear reactors;
- dust from solid rocket motors;
- paint flecks.

Lastly, each of these debris objects is spinning and there is simply no way of knowing their spin rate or direction. Each object will, quite literally, be different.

Since the goal is removing the debris object from space, it is not essential to vaporize it. Instead, a laser pulse can simply give the object enough of a "bump" to cause it to change its orbit, with the goal of placing it on an Earth entry trajectory. This can be accomplished in two ways: laser photon momentum transfer or ablation.

A laser can impart momentum, or motion, to the object by simply reflecting light from it. Light has no rest mass, but it does have momentum. This means that the laser light particles striking the debris object can give it a push as the light reflects from it, in exactly the same way as a solar sail derives thrust from reflecting sunlight. Unfortunately, the push of light is very, very small and it would require a sustained interaction between the laser and the object to give it a significant push.

A laser can also heat up the surface of the debris object to the point at which some of the object's material boils away into space. Much like a rocket, the ablated material ejected from the debris object's surface will cause the debris object to react and move in the opposite direction. If enough material is ablated, then the object will receive a significant bump, or impulse, causing it to change its orbital velocity – both speed and direction. With planning, the laser can ablate enough material to cause the object to be sent to a lower orbit where the atmospheric density is larger; from here, it will then fairly quickly enter the Earth's atmosphere and burn up.

Space lasers sounds fanciful to some, but they are closer to reality than you might think. The US Army is testing solid-state kilowatt-level lasers for use on the battlefield on the Earth. These lasers use electrical power and can be carried around on the back of a conventional semi-tractor trailer truck. It is useful to point out that we can generate power levels sufficient to run these lasers using state-of-the-art solar arrays now routinely flown in space. We also routinely loft spacecraft of roughly the same size as a semi-tractor trailer truck – the Hubble Space Telescope, for example, is 13.3 m long. Granted, today's battlefield lasers are not perfected and they are rather fragile. However, they are not as far away from being flyable in space as one might otherwise imagine.

Studies have also examined the use of even more powerful lasers being built on the ground, shooting up through the Earth's dense atmosphere, and zapping debris. This approach doesn't require the same degree of miniaturization as would be required for space-based lasers, but it would require lasers much more powerful than those currently in the development pipeline. Shooting through the dense atmosphere near the ground to space is much more difficult, for many reasons. Perhaps the biggest problem is the scattering of the laser light: the laser light will scatter from the air and the moisture and dust contained within it, losing power literally every inch of the way [11].

We are, however, a long way from being able to fly such lasers in space or from building them on the ground. For one, a laser capable of destroying orbital debris would also be capable of blinding or damaging (someone else's) satellites and will likely be considered a space weapon. Even if the laser were flown such that the weapon concerns were addressed (jointly operated by several space-faring countries, perhaps?), there would still be the issue of not accidentally missing the target and hitting someone else's satellite by accident. It is going to be not so easy to acquire the millimeter-sized debris object traveling in excess of 8 km/sec and hit it with a beam of light. And, if you miss, then you run the risk of the beam striking another satellite in the line of sight somewhere behind the object being targeted. A ground-based laser has the added complication of requiring that no aircraft be in the area during its operation. Accidentally striking a satellite or aircraft would, by anyone's definition, be a bad day.

Frozen Mist
One of the more creative ideas for removing small debris objects is called "frozen mist". In this approach, a rocket launches a spacecraft containing carbon dioxide or some other super-cooled liquid or gas into orbit near a cluster of small debris objects. The spacecraft positions itself in the path of the debris and sprays the frozen mixture in front of it. The droplets act in the same manner as the atmosphere and slow down the debris objects as they collide; instead of atmospheric drag, this might be called mist drag. Estimates from NASA's Ames Research Center, where the idea originated, indicate that such a mist could de-orbit hundreds of debris objects during each interaction [12].

The upside is that the mist will itself de-orbit rather rapidly and not cause any long-term collision risk in or of itself. The downside is the number of separate launches that would be required to successfully remove the hundreds of thousands of small debris objects now circling the globe.

There are far fewer options for removing small debris from orbit than large and the reason is quite apparent: they are very small, extremely difficult to track, and spread out in orbits all around the Earth. Clearly, something needs to be done about them but practical ideas for doing so are few and far between.

Will we move forward to remove existing debris from space, large or small? Whose responsibility is it to clean up space? A related and most important question is who will pay. NASA, ESA, and most other civilian government space agencies are budget-limited. In other words, if they are asked to spend hundreds of millions of dollars to reduce the amount of debris in space, then they are likely to have to pay for doing so out of existing budgets. What science or exploration mission will they cut to find the money to pay for orbital debris removal?

If the military is asked to pay for removing the debris, then who would believe there wasn't an ulterior motive? After all, one person's debris object, large or small, might be another's secret satellite system. Would the world accept the word of any nation's military that they were cleaning up the space environment, removing satellites at will, and that they wouldn't try to exploit this capability for a military advantage in the future?

Should the taxpayers in any single country be asked to clean up the mess made by another country? Most of the debris in space today was created by the United States and Russia (or the old Soviet Union), with China quickly catching up. Clearly, the primary responsibility for cleaning up the mess should lie with them also.

Someone needs to accept responsibility for reducing the amount of debris, soon, or the problem will only get dramatically worse.

References

[1] Liou, J.-C.; Johnson, N.L.; Hill, N.M. Controlling the Growth of Future LEO Debris Populations with Active Debris Removal. *Acta Astronautica*, **66**(5–6) (2010).

[2] Liou, J.-C.; Johnson, N.L. A Sensitivity Study of the Effectiveness of Active Debris Removal in LEO. *Acta Astronautica*, **64**(2–3) (2009).

[3] Kennedy, F. Orbital Express: Accomplishments and Lessons Learned. *Advances in the Astronautical Sciences*, **131**, 575–586 (2008).

[4] Johnson, L.; Khazanov, G.; Gilchrist, B.; Hoyt, R.; Stone, N.; Lee, D. Space Tethers. *Journal of Space Technology and Science*, **26**(1) (2012).

[5] Pearson, J.; Levin, E.; Oldson, J.; Carroll, J. ElectroDynamic Debris Eliminator (EDDE): Design, Operation, and Ground Support. Proceedings of the Advanced Maui Optical and Space Surveillance Technologies Conference, Wailea, Maui, Hawaii, 14–17 September 2010.

[6] Hoyt, R.P.; Robert, L. Foreword: The Terminator Tether™: Autonomous

Deorbit of LEO Spacecraft for Space Debris Mitigation. 38th Aerospace Sciences Meeting and Exhibit, Reno, NV, 10–13 January 2000.

[7] Johnson, L.; Whorton, M.; Heaton, A.; Pinson, R.; Laue, G.; Adams, C. NanoSail-D: A Solar Sail Demonstration Mission. *Acta Astronautica*, **68**(5), 571–575 (2011).

[8] Muhleman, D.; Hudson, R.; Holdridge, D.; Carpenter, R.; Oslund, K. Observed Solar Pressure Perturbations of Echo I. *Science*, **132**(3438) (1960).

[9] Starke, J.; Bischof, B.; Foth, W.; Gunther, J. ROGER a Potential Orbital Space Debris Removal System, 38th COSPAR Scientific Assembly, Bremen, Germany, 15–18 July 2010.

[10] Foust, J. A Sticky Solution for Grabbing Objects in Space. *MIT Technology Review*, 5 October 2011.

[11] Phipps, C.; Lander, M. What's New for Laser Orbital Debris Removal. Beamed Energy Propulsion: Seventh International Symposium, Ludwigsburg, Germany, 10–14 April 2011.

[12] Kusher, D. Five Ideas to Fight Space Junk. POPSCI online, 13 July 2010, available online at *www.popsci.com/technology/article/2010-07/cluttered-space?page=1*.

15 Harden against Space Radiation (Contributed by Dr James K. Woosley)

Some amount of hardening against space radiation is already incorporated into satellite design. However, as we learn more about the space radiation environment and particularly solar flares, the risk appears to be greater than was first supposed. Future generations of satellites must be shielded or include protective measures against the worst-case radiation loads of the most intense solar flare foreseeable. This hardening will also be helpful in the worst-case event of a nuclear detonation in space.

In order to preserve our space resources in the presence of the worst possible solar storms (and also in the possible case of nuclear war in space), it is necessary to harden our assets to these effects.

Hardening cannot be accomplished effectively in isolation, nor can we ignore the effects of space environments on ground systems due to geomagnetic coupling following an intense solar flare or solar storm. Figure 15.1 illustrates

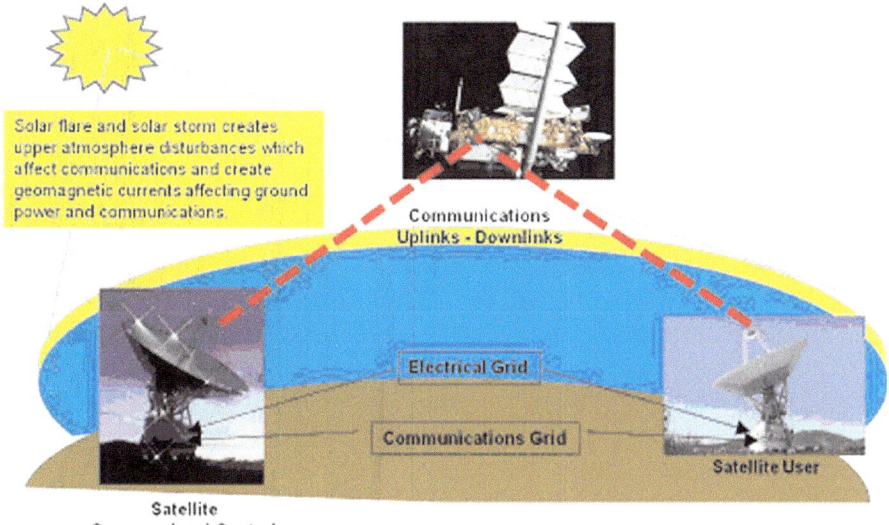

Figure 15.1. Solar storms can affect the electrical and power grid on the ground and temporarily disturb the atmosphere, affecting communications, in addition to inducing radiation effects which can temporarily or permanently impair space systems. (Image courtesy of NASA)

that hardening against space radiation requires a systematic consideration of every element of the space system – including the ground-based command and control and the various ground-based users of the system – which are susceptible to the solar storm.

In the remainder of this chapter, the focus will be on hardening of space systems. There have been numerous studies of the work necessary for concurrent hardening of terrestrial electrical and communication grids, which is also important but is outside the scope of this book.

Hardened design of satellites must consider three factors:

- the prompt effect of the flare X-rays (Figure 15.2), particularly on exposed sensor surfaces and on solar panels;
- the prompt effects of high radiation dose (Figure 15.3), which can cause system degradation and failure due to so-called single event effects;
- the cumulative effect of high radiation dose, which causes the breakdown of electronic components exposed to radiation over time.

The same considerations apply to hardness against nuclear environments, although the nuclear burst radiation in space differs both qualitatively and

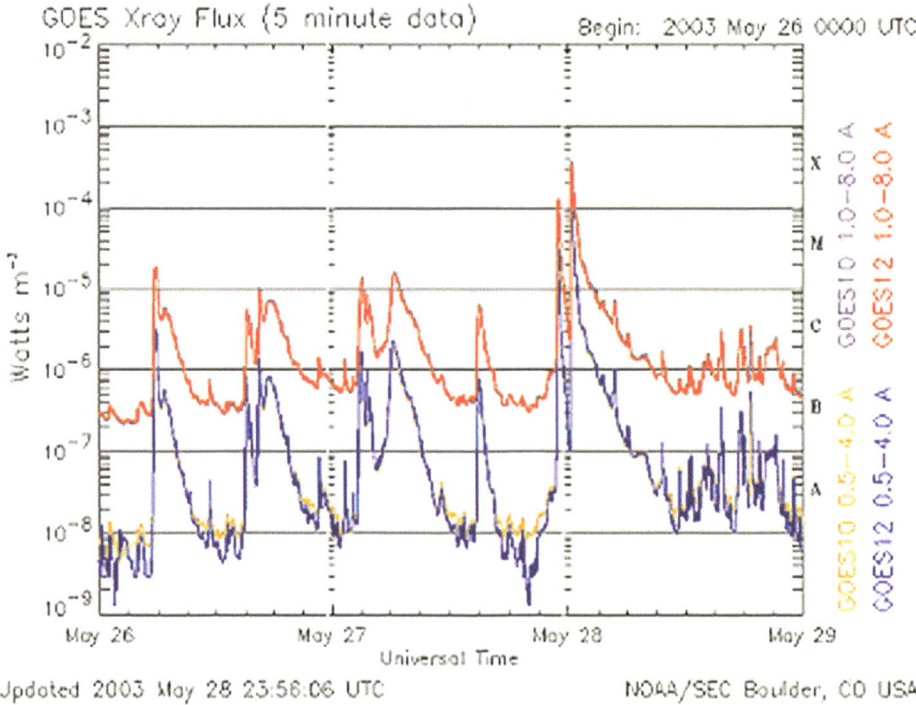

Figure 15.2. GOES X-ray flux, 26–29 May 2003, showing a series of M- and X-class X-ray flares. (Image courtesy of NASA)

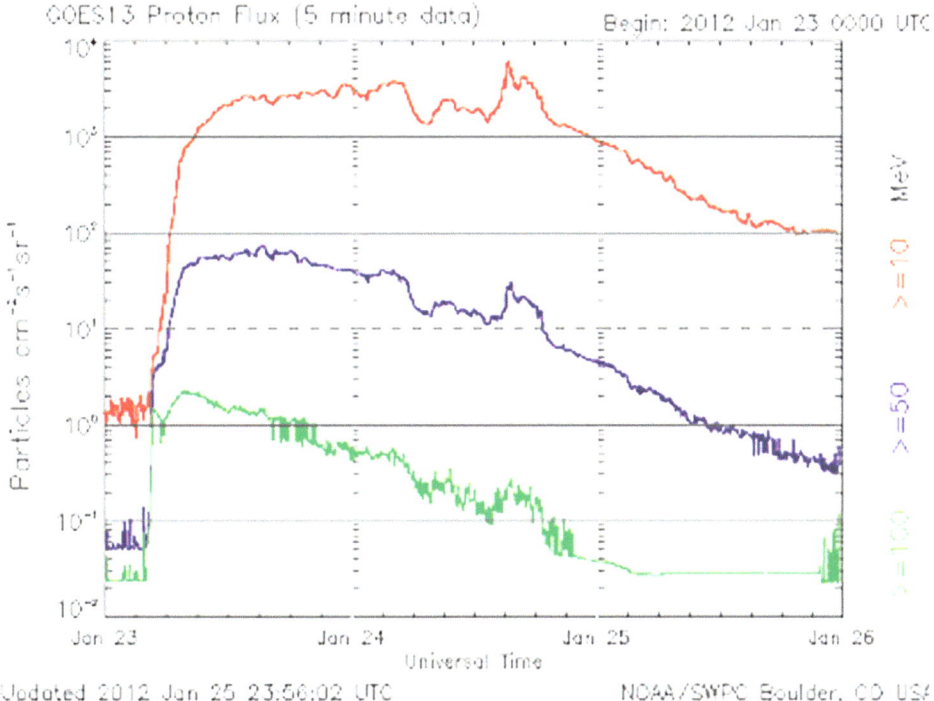

Figure 15.3. GOES 13 Proton Flux for 23–26 January 2012, showing an S-3 radiation storm following an X-class X-ray flare. (Image courtesy of NASA)

quantitatively from solar weather-induced radiation effects – qualitatively because the nuclear ionizing radiation consists primarily of gamma rays and beta rays (electrons) instead of the protons from the solar particle event and coronal mass ejection (CME); and quantitatively because the total energy impulse and energy distribution of each radiation type vary from event to event for both nuclear and solar sources, and more significantly between sources.

For solar flare X-rays (Figure 15.2), the principal concern is the radiation damage imposed on sensitive surfaces such as sensor elements and solar panels. For both types of solid-state devices, radiation damage to the elements and damage due to the induced power surges can occur; there are recorded cases of satellite power systems being damaged or destroyed by solar flare X-rays. And, just as there is a magnetic inflection caused on the ground by the interaction of flare X-rays with the upper atmosphere, at sufficiently high X-ray flux, a similar electromagnetic disturbance can be caused in any exposed equipment. This phenomenon is referred to as System-Generated Electromagnetic Pulse (SGEMP).

Single event effects (or SEE) refers to any disruption of a circuit caused by a single charged particle traversing the circuit, causing ionization and charge separation. The resulting electrical impulse is what causes the disruption. There are numerous types of SEE, depending on the type of circuit and/or the

Table 15.1. Major types of single event effects (SEE) (adapted from *http://radhome.gsfc.nasa.gov/radhome/papers/seeca1.htm*)

SEE type	Description
Single Event Upset (SEU)	The charge induced by the radiation causes a change in a single bit (or a series of bits) of memory lying along the radiation's path. The bit recovers the next time data are written to it, but the results of the affected computation are disrupted
Single Hard Error (SHE)	The radiation transfer permanently damages a bit
Single Event Functional Interrupt (SEFI)	Combinations of bits are disrupted, which causes the circuit or device to stop working effectively, but without hard damage. Recovered by a system reset
Single Event Latch up (SEL)	The circuit is locked up in a high power state and could burn out. Requires a complete power down and restart to recover
Single Event Burnout (SEB)	The self-destruction of a high-power circuit due to overload after the particle passes

component disrupted by the traversing particle. The most significant types are illustrated in Table 15.1. SEE problems are not limited to space systems; at one time, the occurrence of SEE in the electronic controller was suggested as a cause for an automotive brake problem.

Total dose damages are primarily due to progressive damage to circuits exposed to radiation as the individual atoms of the circuit are pushed out of position, which, over time, degrades the efficiency of the circuit.

The principles of establishing radiation-hardened electronics are well established, but add considerably to the cost of space systems. Most NASA programs look to the 1998 NASA Goddard paper, "Emerging Radiation Hardness Assurance (RHA) Issues: A NASA Approach for Space Flight Programs" [1], for guidance; more stringent requirements are followed for military space systems and systems with specified nuclear radiation hardness assurance requirements.

The principles of radiation hardness assurance can be summarized as follows:

- Define the radiation environment in which the space system is required to operate. This definition should be comprehensive and as conservative as possible. (In this case, "conservative" means that the system should be designed to work in the worst possible conditions.)
- Evaluate the effects of the radiation environment on the satellite system. This evaluation has to consider all radiation effects concurrently and has to consider the overall satellite design, since the satellite mass adjacent to the electronics can both shield the electronics from lower-level radiation and create interactions with higher-energy radiation, which makes this more damaging to the electronics.
- Define the requirements for radiation-hardened circuits. This typically involves using the defined radiation environment to estimate the design levels of radiation. Depending on the effects present, the circuit is usually

defined to operate at two to five times the estimated environment, with the actual requirements considering which environments will be tested and how different circuits respond differently to different types of radiation. This will also consider whether the system must remain functional through the enhanced radiation environment or whether it can partially power down and reset/restart after the pulse of radiation subsides.

- Evaluate the independent electronics components for historical test data and determine whether new testing is required.
- Complete the engineering of the system through component qualification and testing and assessing the risk of alternative design approaches.

Most current space system design has come to rely on the extreme solar storms of the Space Age for design guidance. Since the characteristics of a Carrington-type event are not known, we can only guess that the radiation levels that would be encountered are in the range of three to five times greater than anything we have recorded to date. (For a full discussion of the Carrington event, please refer to Chapter 3.) Satellites not specifically designed with this additional margin in mind – which comprise most current commercial satellites – will have degraded performance and will likely fail quickly after a Carrington-type event.

Reference

[1]   LaBel, K.A., et al. Emerging Radiation Hardness Assurance (RHA) Issues: A NASA Approach for Space Flight Programs. *Nuclear Science, IEEE Transactions,* **45**(6), 2727–2736 (1998).

Appendices

A   More on Orbital Debris

Orbital debris-related events are happening just about every day, further increasing the amount of debris in space and enhancing the probability of a coming crisis there. Cataloged here are but a few recent examples of these events. All items from the *NASA Orbital Debris Quarterly News* are in the public domain and provided courtesy of NASA.

From the *NASA Orbital Debris Quarterly News*, Volume 16, Issue 4, October 2012

"New Russian Launch Failure Raises Breakup Concern
The failure of a Russian launch vehicle upper stage on 6 August has led to concerns that the partially-loaded rocket body might explode as two similar stages have done in recent years. Like its sisters, the new Proton Briz-M stage was left stranded in an elliptical orbit with a perigee in low Earth orbit (LEO), where debris from a future explosion could pose a threat to numerous operational spacecraft there. This latest accident also left two fully-fueled spacecraft in orbits like that of their carrier.

The launch malfunction occurred when the Briz-M upper stage (International Designator 2012–044C, U.S. Satellite Number 38746), carrying the Telkom 3 and Express MD2 spacecraft, shut-down shortly after the start of the third of its planned four maneuvers. At the time, the stage was in an orbit of 265 km by 5015 km with an inclination of 49.9 deg. Both spacecraft were later autonomously released from the stage.

The Briz-M stage has a diameter of 4.0 m and a length of 2.65 m and is comprised of two major parts: (1) a core section with a central propellant tank carrying a propellant mass of 5.2 metric tons and the main propulsion system and (2) an auxiliary propellant tank (APT) with an initial propellant mass of 14.6 metric tons and shaped like a donut to encompass the core section. During a normal mission, the APT is separated after the first few burns, and the core section completes the payload delivery task. In the 6 August accident, the APT had not yet been separated, leaving more than 5 metric tons of propellant in the integrated 2.6 metric ton (dry mass) stage.

Concern about a future explosion of the Briz-M stage is based upon the breakups of two Briz-M stages (one in 2007 and one in 2010) that had suffered very similar flight malfunctions. In these two earlier cases, the failures also occurred prior to separation of the APT. The breakups, however, did not take place until 12 and 31 months, respectively, after the failures (ODQN April 2007,

p. 3 and January 2011, pp. 2–3). Although only about 100 large debris from each of these Briz-M stages have been cataloged to date (Figure 2), a much larger number of hazardous debris are believed to stretch from 300 to nearly 29,000 km.

Another Briz-M stage suffered an on-orbit malfunction in August 2011. In that case, the failure took place after the APT was released. This stage contains less than 5 metric tons of propellant and has not yet exploded, although the potential for a severe breakup is believed to still exist. The orbital lifetimes of the 2011 and the 2012 Briz-M stages are estimated to be at least several decades.

At the end of a nominal mission, any residual propellants and pressurants in the central tank are vented into space, according to international orbital debris mitigation guidelines. This passivation process, however, is not guaranteed to function in at least some failure scenarios, as evidenced by the two aforementioned Briz-M breakups.

In addition to the stages themselves, thedisposition of the spacecraft must be addressed. The 2006 and 2011 missions each carried a single, large spacecraft of European manufacture: Arabsat 4A and Express AM4, respectively. After consideration of various disposal options, both spacecraft were sent on controlled, destructive reentries over the Pacific Ocean. This decision eliminated the risk of a future collision by either of the spacecraft with another resident space object, as well as the risk of a debris producing explosion.

In the case of the 2008 failure, the AMC 14 spacecraft, manufactured in the U.S., had sufficient propellant reserves to gradually raise itself and eventually to enter a useful geosynchronous orbit, although one with an inclination of 13 deg. No decision has yet been made regarding the fates of Telkom 3 and Express MD2.

Finally, another variant of the Briz-M, the Rokot Briz-KM has also encountered two failures since 2005. In the first instance, a malfunction prevented the stage and its payload, the European Cryosat, from even entering an Earth orbit. The second mishap occurred in February 2011 and stranded a Russian geodetic satellite, GEO-IK 2, in anelliptical orbit when a second burn of the stage was not executed. This stage, too, is believed to contain a significant amount of residual propellants, which might someday trigger an explosion."

From the *NASA Orbital Debris Quarterly News*, Volume 16, Issue 3, July 2012

"Status of Three Major Debris Clouds
The U.S. Space Surveillance Network continues to catalog debris from the two most prolific events in Earth orbit: the intentional destruction of the Chinese Fengyun-1C spacecraft in January 2007 and the accidental collision of the Russian Cosmos 2251 and the U.S. Iridium 33 spacecraft in February 2009. By 1 July a total of more than 5500 debris had been officially cataloged with these breakups, 90% of which were still in orbit about the Earth. These debris represent 36% of all objects residing in or passing through low Earth orbit, i.e., less than 2000 km altitude.

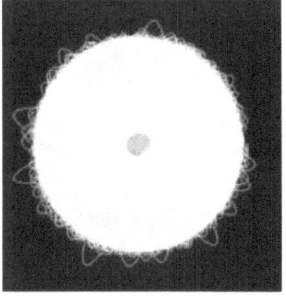
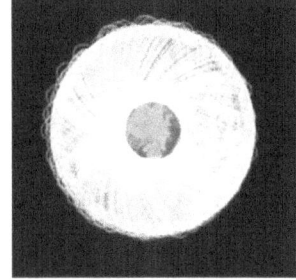
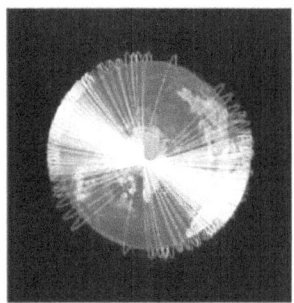

Fengyun-1C debris orbits *Cosmos 2251 debris orbits* *Iridium 33 debris orbits*

Figure A.1

The illustrations above (Figure A.1) clearly indicate that the debris from Fengyun-1C and Cosmos 2251 now completely encircle the planet. Since Iridium 33 was in a nearly polar inclination (86.4 degrees), the orbital planes of its debris are taking longer to diverge as a result of lower differential precession rates."

From the *NASA Orbital Debris Quarterly News*, Volume 16, Issue 2, April 2012

"Chinese Rocket Body Explosions Continue
For the fourth time in 5 years, the third stage of a Chinese Long March 3 (CZ-3) launch vehicle has exploded shortly after delivering its payload to a highly elliptical geosynchronous transfer orbit (GTO). The most recent event occurred on 26 February 2012, within 2 days of the successful launch and release of the vehicle's Beidou G5 navigation satellite.

After lift-off from the Xichang Satellite Launch Center on 24 February 2012, the third stage of a CZ-3C launch vehicle (International Designator 2012-008B, U.S. Satellite Number 38092) carried the Beidou 5G satellite to an orbit of approximately 200 km by 36,000 km with an inclination of 21 degrees. The third stage, with a dry mass of 3 metric tons, a length of 12.4 meters, and a diameter of 3 meters, remained in this GTO orbit, while its payload propelled itself into a geosynchronous orbit. Two days after launch, the U.S. Space Surveillance Network (SSN) detected dozens of debris associated with a severe fragmentation of the stage. To date only one of these debris, believed to be the vehicle equipment bay (VEB), has been officially cataloged.

Following two serious explosions of Long March 4 stages in low Earth orbit in 1990 and 2000, respectively, China established a requirement in 2005 to deplete all residual propellants from orbital stages after payload separations. Such passivation measures have been implemented for the Long March 2 and Long March 4 series launch vehicles, which employ hypergolic propellants in their final stages. However, according to a report from the China National Space Administration (CNSA) in 2011, efforts to design passivation procedures for Long

March 3 third stages, which burn liquid hydrogen and liquid oxygen, had not yet been completed [1]. The same report stated that the typical amounts of residual hydrogen and oxygen were 110 kg and 500 kg, respectively.

The three previous CZ-3 third stage explosions occurred in February 2007, November 2010, and December 2011, and each took place within 2 days of launch. Although 30 new debris were cataloged with the first event, no new debris were cataloged with the other two breakups. In part, this is due to SSN limitations in observing small debris in low inclination, highly elliptical orbits. One of the stages (International Designator 2010-057B, U.S. Satellite Number 37211) reentered the atmosphere in late September 2011. Presumably, all or most of its associated debris have already reentered."

From *NASA Orbital Debris Quarterly News*, Volume 15, Issue 3, July 2011

"Fiftieth Anniversary of First On-Orbit Satellite Fragmentation
On 29 June 1961, a U.S. Ablestar upper stage exploded into nearly 300 large pieces, overwhelming the then official total Earth orbital population of only 54 objects. The 50th anniversary of this seminal event was marked this year with both reflection and optimism: reflection on the more than 200 known satellite fragmentations which followed and optimism that current space vehicle designs and operations will continue to curtail such accidental occurrences in the future.

The Ablestar stage (International Designator 1961-015C [aka 1961-Omicron 3], U.S. Satellite Number 118) lofted the Transit 4A spacecraft along with two smaller scientific satellites, Injun 1 and Solrad 3, into an orbit 880 km by 1000 km with an inclination of 67 degrees. Transit 4A was one of the early members of the first global navigation satellite system and carried the first nuclear power supply into space, the SNAP-3 radioisotopic thermoelectric generator (RTG). Although the piggyback satellites failed to separate from one another, the mission was deemed successful with the Ablestar stage completing all of its required tasks. However, just 77 minutes after orbital insertion, the stage violently came apart, throwing debris across the entire low Earth orbit (LEO) region with some of the fragments reaching to altitudes above 2000 km.

Despite the rudimentary nature of space surveillance sensors in the early 1960s, this satellite breakup remains one of the best documented since the vehicle was being observed with both radio and optical means at the time of the event. From a southeasterly launch from Cape Canaveral, Florida (Figure 1), the Ablestar stage led the Transit 4A and the still-joined Injun 1 and Solrad 3 satellites as the trio passed for the first time over the western United States. At 0608 GMT a Baker-Nunn camera in Organ Pass, New Mexico, clearly photographed the three distinct vehicles, while an electronics van in Downey, California, received a beacon signal from Ablestar (Figure 2). Seconds later, the signal reception ceased, and the image of the Ablestar stage became a blur.

The explosion of an artificial satellite was unprecedented, but within only weeks over 100 debris had been identified. As the capability of the U.S. Space Surveillance Network (SSN) improved over the years, the number of known large debris (>10 cm) from the approximately 600-kg Ablestar gradually grew to reach 293 in June 1992, 31 years after its explosion. [Three of the 296 debris officially cataloged with this event are now known to have originated with other break-up events.] Due to the high altitude of the event, 60% of these debris (176 in all) remain in Earth orbit today (Figure 3).

A thorough investigation into the possible causes for the catastrophic event was immediately undertaken. A preliminary report identified two basic mechanisms which might have caused the stage to break apart: the mixing of the hypergolic propellants prior to break-up or the explosive depressurization of the propellant tanks to allow non-explosive burning of the propellants. At the time of the explosion, the stage contained an estimated 60 kg of fuel (inhibited red fuming nitric acid or IRFNA) and 41 kg of oxidizer (unsymmetrical dimethylhydrazine or UDMH), both under about 320 psia pressure (Figure 4). The stage also carried three helium pressurant tanks and a nitrogen-fed attitude control system.

Four means were found as possible triggers of the above mechanisms:

a) Command destruct system initiation
b) Propulsion system tankage or value leakage or rupture
c) Electrical/electronic system malfunction
d) External heating or particle impact.

A detailed study of the command destruct system found it to be a highly improbable cause of the explosion. A minor meteor shower was underway during 27–30 June that year, but this, too, is viewed as an unlikely root cause, as is an electrical or electronic system failure.

This leaves the propulsion system itself as the likely reason for the breakup. Investigators noted that only a 30-40 psi decrease in the oxidizer tank pressure 'would result in the inversion of oxidizer/fuel tank intermediate bulkhead. This in turn would result in rupture of the bulkhead and mixing of the residual hypergolic propellants.' The report went on to note five different ways for the oxidizer tank to lose pressure.

Perhaps the most telling finding was that on all previous Ablestar missions the fuel tank had been vented as part of the spacecraft separation process. However, due to concerns about possible contamination of venting fuel on the payloads, for the Transit 4A mission a separate helium-based retro system had been employed, and the fuel tank was left pressurized.

Since a single root cause could not be absolutely determined, a number of countermeasures were recommended for subsequent Ablestar missions, including the venting of the fuel tank and the disabling of the range safety system after release by the range safety officer.

Unfortunately, it was not until the 1980s and 1990s that passivation of all launch vehicle orbital stages became recognized as necessary to combat a number of intensive accidental explosions by stages from the U.S., Russia, China,

India, the Ukraine, and the European Space Agency. Today, stage passivation measures are common and are explicitly recommended by several national space agencies, by the Inter-Agency Debris Coordination Committee (IADC), and by the United Nations."

From *NASA Orbital Debris Quarterly News*, Volume 15, Issue 1, January 2011

"International Space Station Avoids Debris from Old NASA Satellite
Once again in October 2010, the International Space Station (ISS) was forced to maneuver to avoid a potential collision with large orbital debris. Such maneuvers now occur about once per year. This time the threatening object was a piece of debris which had come off a 19-year-old NASA scientific spacecraft only one month earlier.

Since its decommissioning in late 2005, NASA's Upper Atmosphere Research Satellite (UARS) had been gradually falling back to Earth, and by late September 2010 the 5.7-metricton spacecraft was in an orbit of 335 km by 415 km with an inclination of 57.0 degrees, when the U.S. Space Surveillance Network (SSN) discovered that a fragment had separated from the vehicle. This was not unprecedented since four other pieces of debris had separated previously in November 2007 (Orbital Debris Quarterly News, January 2008, p. 1). The reasons for these releases remain unknown.

The new piece of debris, later cataloged as International Designator 1991-063G and U.S. Satellite Number 37195, was ejected with some force, resulting in an initial orbit of 375 km by 425 km, which was actually higher than the orbit of UARS. With its much greater area-to-mass ratio, the fragment began falling back to Earth much more rapidly than UARS itself, finally reentering the atmosphere on 4 November, after an orbital lifetime of only 6 weeks. For comparison, UARS is not expected to reenter until the summer of 2011.

Operations Center (JSpOC) and NASA's Houston Mission Control Center calculated that the fragment from UARS would come unacceptably close to the ISS the following day, 26 October. The probability of collision was assessed to exceed 1 in 10,000, which is the nominal threshold for executing a collision avoidance maneuver. At the time, the orbit of the 370-metric-ton ISS was approximately 350 km by 360 km, and the orbit of the debris was 325 km by 360 km.

After updates of the respective orbits, new assessments confirmed a close approach which would violate standing safety protocols. Consequently, a decision was made to conduct a small, posigrade maneuver (i.e., + 0.4 m/s) a little more than 2 hours before the anticipated flyby. The collision avoidance maneuver was successfully performed by the Progress M-07M logistics vehicle which had docked at the aft port of the ISS Zvezda module on 12 September."

From *NASA Orbital Debris Quarterly News*, Volume 15, Issue 1, January 2011

"New Satellite Fragmentations Add to Debris Population
The U.S. Space Surveillance Network (SSN) detected four new satellite fragmentations during October and November, three involving relatively young launch vehicle stage components and one associated with an old meteorological spacecraft. Fortunately, none of the events appears to have created large amounts of long-lived debris.

On 14 March 2008 the upper stage of a Russian Proton launch vehicle malfunctioned part way through the second of three planned burns designed to place a commercial spacecraft, AMC-14, into a geosynchronous transfer orbit. Although the spacecraft eventually limped into a geosynchronous orbit, albeit an inclined one, with its own propulsion system, the Briz-M upper stage was stranded in a highly elliptical orbit with a significant amount of residual propellant.

The stage (International Designator 2008- 011B, U.S. Satellite Number 32709) remained dormant until 13 October 2010, more than two and a half years after launch, when it experienced an apparently minor fragmentation. At the time of the event, the stage was in an orbit of 645 km by 26,565 km with an inclination of 48.9 degrees. More than 30 debris from the stage have been provisionally identified by the SSN with orbital periods ranging from 430 to more than 540 minutes. However, to date only eight debris have been officially cataloged. Due to the nature of this highly elliptical orbit, small debris are difficult for the SSN to detect and to track. In February 2007 another Briz-M, which had also failed in its delivery mission, exploded into an estimated 1000+ large fragments (Orbital Debris Quarterly News, April 2007, p.3), although only 92 debris have so far been officially cataloged.

Less than 3 weeks after the fragmentation of the Russian upper stage, a newly launched Chinese launch vehicle stage released dozens of debris for unknown reasons. Beidou G4, the latest addition to China's global navigation satellite system, was launched on 1 November by a Long March 3C rocket. The launch vehicle successfully delivered its payload into a geosynchronous transfer orbit of 160 km by 35,780 km with an inclination of 20.5 degrees, but within a few hours of launch the SSN detected more than 50 debris associated with the final stage of the vehicle (International Designator 2010-057B, U.S. Satellite Number 37211).

This event was reminiscent of the February 2007 breakup of a Long March 3 upper stage carrying another Beidou spacecraft. In that case the debris, which also were released very soon after launch, were initially believed to be associated with the spacecraft, which did encounter early system problems (Orbital Debris Quarterly News, April 2007, p.3). However, later analysis confirmed the source of the debris was the final stage of the launch vehicle.

The third fragmentation event of the fourth quarter of 2010 originated from a 22-year-old U.S. meteorological satellite. Launched in late 1988, the NOAA-11 spacecraft (International Designator 1988-089A, U.S. Satellite Number 19531) operated for more than 15 years before being retired and decommissioned in June 2004. On 24 November 2010, two fragments were ejected from the spacecraft with moderate velocities, one to a higher orbit and one to a lower mean altitude.

At the time, NOAA-11 was in an orbit of 835 km by 850 km with an inclination of 98.8 degrees. The two debris were found in orbits of 815 km by 850 km and 840 km by 860 km and have been cataloged as U.S. Satellite Numbers 37241 and 37242, respectively. Three previous NOAA spacecraft (NOAA-6, NOAA-7, and NOAA-10) are known to have released small amounts of debris, ranging from three to eight, at least a dozen years after launch. All four NOAA spacecraft were part of the TIROS-N series and were launched between 1979 and 1988. The reason for these minor fragmentations remains unknown.

The final satellite breakup of the year occurred on 23 December and involved a piece of launch debris from Japan's IGS 4A and IGS 4B mission in February 2007. Officially known as 2007-005E (U.S. Satellite Number 30590), the object was the only one of 12 launch vehicle debris associated with the mission to still be in orbit nearly 4 years after launch. At the time of the event, the object was in an orbit of approximately 430 km by 440 km with an inclination of 97.3 degrees. The U.S. Space Surveillance Network initially detected less than 10 new debris, of which 3 were cataloged with U.S. satellite numbers 37261–37263 five days after the event. The nature of the object and, hence, the potential cause of its fragmentation, are unknown at this time. Due to the relatively low altitude of the breakup, the newly created debris will be short-lived."

From *NASA Orbital Debris Quarterly News*, Volume 14, Issue 3, July 2010

"Drifting in GEO
A major malfunction on 5 April resulted in a sudden and complete loss of control of the Galaxy 15 spacecraft (2005-041A, U.S. Satellite Number 28884). Originally stationed over the equator at 133 W, the approximately one-ton spacecraft began a very slow drift eastward and, if control over the vehicle cannot be restored, will likely enter a long-period oscillation orbit around the geopotential stable point at 105 W, moving from 133 W to 77 W and back again.

Although Galaxy 15 will pass close to other operational spacecraft in the Western Hemisphere, the U.S. Space Surveillance Network is closely watching its movement and will advise operators of other spacecraft should a potentially hazardous conjunction, i.e., a pass of less than 5 km between two spacecraft, be forecast. In such an event the operator of the other spacecraft will have sufficient warning to perform a collision avoidance maneuver, if warranted.

Galaxy 15 joins a large number (>150) of other derelict spacecraft and launch vehicle orbital stages that are drifting back and forth in the geosynchronous (GEO) region. Last year three Russian spacecraft failed to perform any disposal maneuvers, which are designed to keep the vehicles at least 200 km above the

GEO altitude of 35,786 km. In 2008, two other spacecraft, one U.S. and one Russian, also failed to maneuver out of GEO at their end of mission.

Of more immediate concern to other GEO spacecraft operators was the threat of radio interference from the still functioning Galaxy 15 payload. Several attempts to turn-off Galaxy 15's transmitters failed. However, a loss of attitude control is expected to lead to a shutdown of the transmitters later this year."

From *NASA Orbital Debris Quarterly News*, Volume 14, Issue 2, April 2010

"Update on Three Major Debris Clouds
The first quarter of 2010 marked the third anniversary of the intentional destruction of the Chinese Fengyun-1C spacecraft and the first anniversary of the accidental collision of the U.S. Iridium 33 and Russian Cosmos 2251 spacecraft. The cataloged debris from these three hypervelocity fragmentations now represents an increase in the low Earth orbit (LEO) satellite population of more than 60% (Figure 1).

The total number of debris cataloged by the U.S. Space Surveillance Network (SSN) from Fengyun-1C has continued to grow and had reached 2841 by the end of March 2010, of which less than 85 had reentered. Moreover, more than 500 additional debris were being tracked by the SSN and were awaiting formal cataloging. Meanwhile, the known large debris from the Iridium-Cosmos collision also has increased. The number of cataloged debris from Iridium 33 and Cosmos 2251 now stands at 1228 and 512, respectively, for a total of 1740. About 400 additional debris have also been identified for future cataloging.

Therefore, the combined cataloged population from these two events, less those debris which have already reentered, is more than 4400. These debris are concentrated in the heart of LEO but spread across the entire region (Figure 2). However, the total number of large debris known to still be in orbit is approximately 5500. For debris as small as 1 cm the total number from these three fragmentations alone is more than 250,000."

From *NASA Orbital Debris Quarterly News*, Volume 14, Issue 1, January 2010

"Avoiding Satellite Collisions in 2009
All NASA programs and projects operating maneuverable spacecraft in low Earth orbits (LEO) or in Geosynchronous Earth Orbits (GEO) are required to have periodic conjunction assessments performed for the purpose of avoiding collisions with other known resident space objects. These conjunction assessments are conducted by the Joint Space Operations Center (JSpOC) of the U.S. Strategic Command at Vandenberg Air Force Base in California. For the International Space Station and the Space Shuttle, these assessments are typically

updated three times per day. For robotic satellites, which normally operate at higher altitudes where atmospheric drag effects are less pronounced, the assessments are updated daily, on average.

The conjunction assessment alert messages from JSpOC identify the object which is expected to come near the NASA spacecraft along with information on the predicted time and distance of closest approach, as well as the uncertainty associated with the prediction. Typically, alert messages are issued if the calculated miss-distance is within a few kilometers of the NASA spacecraft. While the sensors of the U.S. Space Surveillance Network are tasked to collect additional tracking data to refine the close approach prediction, NASA specialists compute the actual probability of collision. In the case of human space flight, collision avoidance maneuvers are normally conducted if the risk of collision is greater than 1 in 10,000. In general, robotic spacecraft accept higher levels of risk, i.e., on the order of 1 in 1,000.

Most alert messages do not result in collision avoidance maneuvers. Often, a recomputation of the conjunction assessment with updated tracking data and a shorter propagation period (i.e., time to the encounter) will reveal a more distant miss-distance and a lower risk of collision. It is not uncommon for collision avoidance maneuvers to be planned but canceled when new assessments are completed. For example, the International Space Station was prepared to conduct a collision avoidance maneuver on 17 March to evade a piece of debris from a former Soviet satellite which exploded in 1981. A later conjunction assessment revealed that the maneuver was not necessary. The debris reentered the Earth's atmosphere on 4 April, no longer posing a threat to the International Space Station or other satellites.

During 2009 conjunction assessments led to eight collision avoidance maneuvers by NASA spacecraft, in addition to a collision avoidance maneuver of a French satellite operating in concert with NASA Earth observation satellites (Table 1). Only two of the maneuvers involved close approaches by intact vehicles (one a spacecraft and one a rocket body). The other maneuvers were needed to avoid collisions with smaller debris, including twice with debris from

Table 1 Collision avoidance maneuvers in 2009

Spacecraft	Maneuver Date	Object Avoided
TDRS 3	January 27	Proton rocket body
ISS	March 22	CZ-4 rocket body debris
Cloudsat	April 23	Cosmos 2251 debris
EO-1	May 11	Zenit rocket body debris
ISS	July 17	Proton rocket body debris
Space Shuttle	September 10	ISS debris
PARASOL	September 29	Fengyun-1C debris
Aqua	November 25	Fengyun-1C debris
Landsat 7	December 11	Formosat 3D

the Chinese anti-satellite test of 2007 and once with debris from the collision of the Iridium 33 and the Cosmos 2251 satellites in February of 2009.

On a separate occasion in March 2009, the crew of the International Space Station had to retreat temporarily into their Soyuz return spacecraft when debris from a U.S. upper stage were projected to make a close approach (ODQN, Vol. 13, Issue 2, p. 3). The elliptical nature of the debris' orbit (about 145 km by 4230 km) contributed to a late notification of the conjunction, leaving too little time to prepare for a collision avoidance maneuver."

From *NASA Orbital Debris Quarterly News*, Volume 13, Issue 1, January 2009

"*New Debris Seen from Decommissioned Satellite with Nuclear Power Source*
A 21-year-old satellite containing a dormant nuclear reactor was the source of an unexpected debris cloud in early July 2008. Launched by the former Soviet Union in February 1987, Cosmos 1818 (International Designator 1987-011A, U.S. Satellite Number 17369) was the first of two vehicles designed to test a new, more advanced nuclear power supply in low Earth orbit. Dozens of small particles were released during the still unexplained debris generation event.

Cosmos 1818 and its sister spacecraft, Cosmos 1867 (Figure 1), carried a thermionic nuclear power supply, in contrast to the simpler, thermoelectric nuclear device which provided energy to the well-known RORSATs (Radar Ocean Reconnaissance Satellites) during the 1970s and 1980s. The most infamous RORSAT was Cosmos 954, which rained radioactive debris over Canada in 1978 after suffering a loss of control malfunction.

Unlike their RORSAT cousins, which operated in very low orbits near 250 km, Cosmos 1818 and Cosmos 1867 were directly inserted into orbits near 800 km, eliminating any threat of premature reentry. According to Russian reports, the nuclear reactors on Cosmos 1818 and Cosmos 1867 functioned for approximately 5 and 11 months, respectively. For the next two decades, the two inactive spacecraft circled the Earth without significant incident.

Following the fragmentation event on or about 4 July 2008, the U.S. Space Surveillance Network was able to produce orbital data on 30 small debris (Figure 2). The majority of these debris were ejected in a posigrade direction with velocities of less than 15 meters per second, suggesting a relatively low energy event. From radar detections, a larger number of very small debris appear to have also been released, but routine tracking of these debris has proven difficult.

Special observations of a few of the debris revealed characteristics generally indicative of metallic spheres. Cosmos 1818 employed sodium potassium (NaK) as a coolant for its reactor, as did the older RORSATs. Although the post-Cosmos 954 RORSATs were known for releasing sign amounts of NaK droplets after reaching their disposal orbits (Kessler et al., 1997), Cosmos 1818 and Cosmos 1867 did not follow this precedent. Much of the NaK within Cosmos 1818 probably was in a solid state at the time of the debris generation event. However, some NaK present in the radiator coolant tubes might have reached a temporary liquid state, particularly when the spacecraft was exposed to direct solar illumination. A breach in a coolant tube (for example, due to long-term thermal stress) at such a time could have resulted in the release of NaK droplets. Alternatively, the hyper-velocity impact of a small particle might have generated sufficient heat to melt some of the NaK, which then would have formed spheres with metallic properties.

Additional analysis of the debris is underway in hopes of providing new insights into the nature of the objects and the possible cause of their origin. To date, no similar debris generation by Cosmos 1867 has been observed."

B The Kessler Effect as Originally Described

Don Kessler's "Kessler Effect" paper from the *Journal of Geophysical Research* is provided below. Because Dr Kessler wrote the paper as part of his official NASA duties, the paper is in the public domain and no copyright can be asserted.

VOL. 83, NO. A6   JOURNAL OF GEOPHYSICAL RESEARCH   JUNE 1, 1978

Collision Frequency of Artificial Satellites: The Creation of a Debris Belt

DONALD J. KESSLER AND BURTON G. COUR-PALAIS

NASA Johnson Space Center, Houston, Texas 77058

As the number of artificial satellites in earth orbit increases, the probability of collisions between satellites also increases. Satellite collisions would produce orbiting fragments, each of which would increase the probability of further collisions, leading to the growth of a belt of debris around the earth. This process parallels certain theories concerning the growth of the asteroid belt. The debris flux in such an earth-orbiting belt could exceed the natural meteoroid flux, affecting future spacecraft designs. A mathematical model was used to predict the rate at which such a belt might form. Under certain conditions the belt could begin to form within this century and could be a significant problem during the next century. The possibility that numerous unobserved fragments already exist from spacecraft explosions would decrease this time interval. However, early implementation of specialized launch constraints and operational procedures could significantly delay the formation of the belt.

INTRODUCTION

Since the beginning of the space age, thousands of satellites have been placed in earth orbit by various nations. These satellites may be grouped into three categories: payloads, rocket motors, and debris associated with the launch or breakup of a particular payload or rocket; most satellites fall into the last category. Because many of these satellites are in orbits which cross one another, there is a finite probability of collisions between them. Satellite collisions will produce a number of fragments, some of which may be capable of fragmenting another satellite upon collision, creating even more fragments. The result would be an exponential increase in the number of objects with time, creating a belt of debris around the earth.

This process of mutual collisions is thought to have been responsible for creating most of the asteroids from larger planetlike bodies. The time scale in which this process is taking place in the asteroid belt is of the order of billions of years. A much shorter time scale in earth orbit is suggested by the much smaller volume of space occupied by earth-orbiting satellites compared to the volume of space occupied by the asteroids.

Conceivably, a significant number of small satellite fragments already exist in earth orbit. Fragments which are undetected by radar are likely to have been produced from 'killer satellite' tests and the accidental explosions of rocket motors. Although some work has already been completed to estimate the number of these fragments, further investigations in this area are still required.

This paper will determine possible time scales for the growth of a 'debris belt' from collision fragments and will predict some of the possible consequences of continued unrestrained launch activities. This will be accomplished by applying techniques formerly developed for studying the evolution of the asteroid belt. A model describing the flux from the known earth-orbiting satellites will first be developed. The results from this model will then be extrapolated in time to predict the collision frequency between satellites. The hypervelocity impact phenomena will then be examined to predict the debris flux resulting from collisions. Other sources and sinks for debris will be discussed, and the effects of atmospheric drag will be predicted. These results will be applied to design requirements for three types of space missions in the future. The potential, or upper limit, debris flux will then be discussed. Although further studies are recommended, the conclusion is

reached by this study that over the next few decades a significant amount of debris could be generated by collisions, affecting future spacecraft designs.

SATELLITE ENVIRONMENT MODEL

A model describing the environment resulting from orbiting satellites was constructed by first calculating the spatial density (average number of satellites per unit volume) as a function of distance from the earth and geocentric latitude. Flux (number of impacts per unit area per unit time) was then related to spatial density through the relative impact velocities. This technique was also used to model the collision frequency in the asteroid belt [*Kessler*, 1971].

Orbital perturbations can be expected to cause the orbital argument of perihelion and right ascension of ascending node to change fairly rapidly, causing these two distributions to be nearly random. This randomness was observed [*Brooks et al.,* 1975] and led to a uniform distribution in the spatial density as a function of geocentric longitude. The model was thus reduced to determining the spatial density S as a function of distance from the surface of the earth R and geocentric latitude β. To construct the model, volume elements were defined as $\Delta R = 50$ km and $\Delta \beta = 3°$. The spatial density in each of these volume elements was found by calculating the probability of finding each satellite in a particular volume element and then summing these probabilities. Spatial density is then this sum divided by the volume of the volume element.

The April 30, 1976, Satellite Situation Report [*NASA,* 1976] contains a total of 3866 satellites, indicating that as of this date a total of 3866 were being tracked, most by radar in low earth orbit. The Satellite Situation Report is compiled from data provided by the Space Defense Center (SDC) and represents the most complete data available. Even so, it was found to be significantly incomplete by a test performed in 1976 [*Hendren and Anderson,* 1976], especially at altitudes below 500 km. In addition, the drop in radar sensitivity for objects smaller than 10 cm [*Brooks et al.,* 1975] and for objects at higher altitudes produces another bias in these data. Such deficiencies caused the calculated spatial densities to be too low; the implications of this result will be discussed in later sections. Since only a statistical solution was required, sufficient accuracy was maintained by performing all calculations with a random sample consisting of 125 of the 3866 satellites.

The resulting spatial densities are given in Figures 1 and 2. Figure 1 shows the spatial density as a function of distance from the earth (averaged over latitude), while Figure 2 shows

Paper number 8A0210.

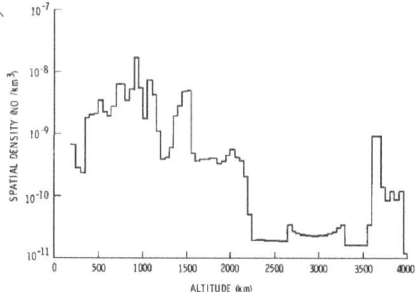

Fig. 1. Current distribution of satellites in earth orbit as observed by radar. A total of 3866 satellites are in the April 1976 catalog and are represented here.

the spatial density as a function of latitude for a few selected distances. Of particular note is that most of the satellites were found within about 2000 km of the earth, the peak density being found at about 900 km. Significant peaks were found at 1500 and 3700 km. In latitude, significant peaks were found between 30° and 35° and at >60°. For most latitudes the spatial density was found to be well within a factor of 2 of the average for that distance, although specific peaks were a factor of 3 or 4 from the average.

As a note of interest at this point, the impact rate on a particular spacecraft, dI/dt, can be quickly approximated from Figure 1 by using the equation

$$dI/dt = S\bar{V}_s A_c \qquad (1)$$

where $\bar{V}_s$ is an average relative velocity, A_c is the cross-sectioned area of the spacecraft, and I is the total number of impacts with the spacecraft at time t. As will be shown later, $\bar{V}_s \simeq 7$ km/s. Therefore for a space station of 50-m radius, at 500-km altitude ($S \simeq 2.8 \times 10^{-9}$/km³ and $A_c = 7.9 \times 10^{-3}$ km²) the impact rate is 1.5×10^{-10}/s, or 4.9×10^{-3}/yr. This compares to about 3×10^{-3}/yr found by *McCarter* [1972] using a 1971 Satellite Situation Report containing 1805 objects. This impact rate is only slightly dependent on the space station orbital inclination: for inclinations less than 90° the rate varies by less than a factor of 2.

The collision rate between all satellites, dC/dt, is given by

$$dC/dt = \frac{1}{2}\int S^2 \bar{V}_s \bar{A}_{cc}\, dU \qquad (2)$$

where C is the number of collisions between satellites, $\bar{A}_{cc}$ is an average collision cross-sectional area of the satellites, and dU is an element of volume. Thus both an average velocity and a collision cross-sectional area are required. These distributions and their resulting averages will be discussed now.

VELOCITY DISTRIBUTION

The velocity distribution in each volume element was calculated by computing the relative velocity between each of the satellites in the random sample and then weighting the number having this velocity by the probability of finding the two satellites in the volume element. Two average velocities were found from these distributions: the average relative velocity $\bar{V}_s$ and the average collision velocity $\bar{V}_c$. To illustrate the differ-

ence between these averages, assume that there are three objects in a unit volume having velocities of 1, 2, and 3 km/s relative to one another. The average relative velocity would be 2 km/s. However, since the probability of collision is proportional to the relative velocity, for every collision at 1 km/s there will be two collisions at 2 km/s and three collisions at 3 km/s. The average collision velocity is thus 2.33 km/s. The average relative velocity can be shown to be the proper average to transform spatial density to flux and collision frequency, while the average collision velocity describes the average velocity at which objects would be observed to collide.

The results of these calculations for altitudes less than 2000 km were that $\bar{V}_s \simeq 7$ km/s and $\bar{V}_c \simeq 10$/km/s; these averages were found to be nearly independent of the volume element. At latitudes where the spatial density was large these average velocities were sometimes slightly smaller (by 1 or 2 km/s); however, this effect was small, so that $\bar{V}_s$ and $\bar{V}_c$ could be considered constant over all space below 2000 km.

SIZE DISTRIBUTION

The size distribution of satellites was obtained from the radar cross section measurements. The data in Figure 3 were obtained during a 12-hour test of the perimeter acquisition radar (PAR) on July 31, 1976 [*Hendren and Anderson*, 1976]. The purpose of this test was to determine PAR's capability to detect and track objects that were not in the Space Defense Center catalog. The results of the test concluded that the SDC catalog was 18% deficient, and the implications of these results will be discussed later. However, the PAR data have other applications, since they represent a 'point-in-time' sampling of the satellite data and were gathered by a single instrument. Thus these data were used to analyze the satellite size distribution.

In an individual case the physical cross section A_c of an object may be orders of magnitude different from its radar cross section σ [*Ruck et al.*, 1970]. However, the difference may be small (less than a factor of 2) when an average physical cross section is compared to an average radar cross section

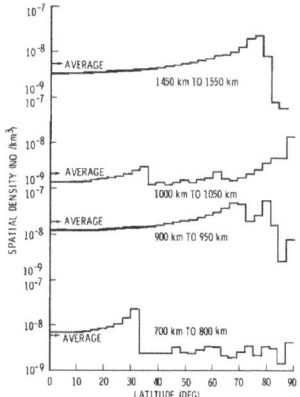

Fig. 2. Latitude variation in spatial density at selected altitudes. At most latitudes the spatial density is within a factor of 2 of the average at that distance.

resulting from the observation of many objects. But when the linear dimensions of an object become small compared to the wavelength, Rayleigh scattering causes the radar cross section to be much smaller than the physical cross section. This drop in radar sensitivity was at least partially responsible for the drop in number for sizes smaller than 0.03 m² shown in Figure 3. Hence the assumption was made that for $\sigma \geq 0.02$ m², $A_c = \sigma$. For $\sigma < 0.02$ m², $A_c \simeq k\sigma$, where k approaches 6 at the smaller values of σ in Figure 3. This correction for Rayleigh scattering produced only minor changes in the overall collision process.

Also plotted in Figure 3 is the distribution of total area of satellites. This distribution is of interest, since collision probabilities are related to area. The distribution suggests that most collisions will involve objects between the sizes of 1- to 100-m² cross section. The average cross section $\bar{A}_c = 1.5$ m² was found by dividing the integral of the area curve by the integral of the number curve.

However, the average collision cross section $\bar{A}_{cc}$ must include the finite size of both objects. Collision cross section is related to the physical cross section of two objects by

$$A_{cc} = (A_{ci}^{1/2} + A_{cj}^{1/2})^2 \qquad (3)$$

where A_{ci} and A_{cj} are the physical cross sections of objects i and j, respectively. The collision cross section was calculated between each of the sizes in Figure 3, weighted according to the probability of those sizes colliding, and then averaged. The results were that $\bar{A}_{cc} \simeq 4$ m² and will be assumed to be independent of the volume element.

When the data in Figure 3 were compared with the SDC catalog data, the PAR radar cross sections were found to be about 50% less, on the average, than those of the SDC. This indicates an additional calibration uncertainty between the two radar systems. When combined with the uncertainty between radar and physical cross sections, the value of $\bar{A}_{cc}$ may range from about 2 m² to 12 m².

Thus (2) was integrated over the space between 150-km and 4000-km altitude by using $\bar{V}_s = 7$ km/s and $\bar{A}_{cc} = 4$ m² to give a collision rate of 0.013 collisions/yr, but this rate could range

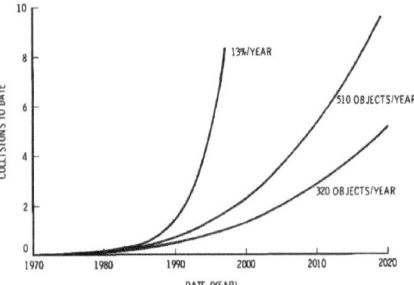

Fig. 4. Total collisions by the given date under various growth assumptions. The first collision is expected between 1989 and 1997.

from 0.007 to 0.039 collisions/yr, owing to the uncertainty in the relationship between the real and radar cross sections. These collision rates could be significantly higher if a large number of unobserved satellites exist.

EXTRAPOLATION INTO THE FUTURE

Between 1968 and 1974 the net number of trackable objects in space increased at the rate of about 320 objects/yr [*Brooks et al.*, 1975]. From 1975 to the present that rate increased to 510/yr [*NASA*, 1974, 1975, 1976]. The period from 1966 to the present could be summarized by an increase of 13%/yr.

If it is assumed that the same pattern of debris buildup continues, then the number of collisions C by time t is found by integrating (2) over time, where the spatial density is then a function of time and proportional to the number of objects in space.

Figure 4 shows the results of such an integration, using the three different buildup rates. Note that under the more conservative assumption (320 objects/yr) the first collision would be expected around 1997. However, at a growth rate of 13%/yr this collision would occur around 1989. If the average collision cross section is overestimated by a factor of 2, the first collision could be as late as the year 2005, while an underestimation by a factor of 3 results in a first collision between 1985 and 1990 under any of the growth assumptions. The presence of unobserved satellites would move these dates even closer to the present. Thus unless significant changes are made in the method of placing objects into space, fragments from intercollisions will probably become a source of additional space debris by the year 2000, perhaps much earlier. The significance of this new source is seen by taking a closer look at the hypervelocity impact phenomenon.

HYPERVELOCITY IMPACTS

The average impact velocity of 10 km/s ensures that almost all of the earth-orbiting objects will exhibit hypervelocity impact characteristics when they collide. Both objects will be subjected to very high instantaneous pressures ($\geq 10^{12}$ dyn/cm²), the strong shock waves causing melting and possible vaporization in the immediate region of the impact. A crater, or hole, will be formed, the molten ejected mass coalescing into more or less spherical particles. In addition, the shock waves, particle fragments hitting other surfaces, and vapor pressure may cause fragmentation outside the cratered region,

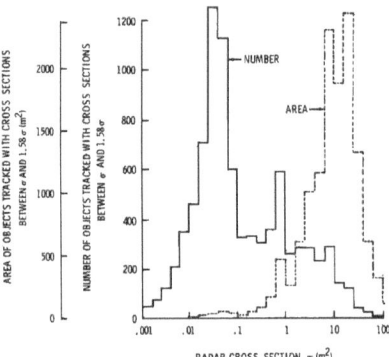

Fig. 3. Size distribution of earth-orbiting satellites observed by radar. The largest number of satellites have a radar cross section of about 0.04 m², while the largest area contribution is around a radar cross section of 10 m².

possibly resulting in the catastrophic disruption of both objects. This process has been studied for some time, mostly from the standpoint of protection of spacecraft from meteoroids, crater formation on the moon, and fragmentation of rocks on the lunar surface or in the asteroid belt. Because of the parallel between the potential formation of an earth-orbiting debris belt and the hypothetical formation of the asteroid belt, and because of the availability of data concerning impacts into solid, homogeneous objects, these data will be discussed first.

IMPACTS INTO SOLID STRUCTURES AND BASALT

Hypervelocity impacts into solid structures can be divided into two groups: catastrophic and noncatastrophic. A noncatastrophic collision results from the collision of two masses M_1 and M_2, where M_1 is much smaller than M_2 by an amount

$$M_2 > \Gamma' M_1 \qquad (4)$$

where Γ' is a function of the impact velocity and the structure and materials of M_1 and M_2. In noncatastrophic collisions, only M_1 is destroyed, and a crater is produced in M_2, ejecting a total mass of M_e, which may be expressed as

$$M_e = \Gamma M_1 \qquad (5)$$

where Γ is also a function of the impact velocity and the structure and materials of M_1 and M_2.

If M_1 is larger than the amount given in (4), then not only is a crater produced in M_2, but the entire structure of M_2 begins to fragment. This process is referred to as a catastrophic collision. These additional fragments are usually larger than the fragments from the crater and are ejected at a much slower velocity. The mass ejected from a catastrophic collision is

$$M_e = M_1 + M_2 \qquad (6)$$

The ejected mass has also been shown to be proportional to the impact kinetic energy [Moore et al., 1965; Dohnanyi, 1971]. Thus the values for Γ and Γ' will vary as V^2. At 10 km/s the values of Γ and Γ' for basalt are 500 and 25,000, respectively [Dohnanyi, 1971]. That is, for basalt, if M_2 is greater than 25,000 times M_1, then the ejected mass resulting from a collision between M_1 and M_2 at 10 km/s is 500 times M_1. If M_2 is less than 25,000 times M_1, then the collision is catastrophic. These results are summarized in Table 1, along with the results for glass and 1100-0 aluminum (low strength, high ductility). The glass and aluminum tests were performed by the Ames Research Center and the General Motors Defense Research Laboratories for the Johnson Space Center.

The number of small fragments of mass M and larger ejected from a noncatastrophic collision can be expressed as

$$N = K(M/M_e)^\eta \qquad (7)$$

TABLE 1. Hypervelocity Impact Parameters

Material	Γ'	Γ
Basalt	25,000	500
Glass	120,000	2,000
1100-0 aluminum	2,600	130
Spacecraft structure	>115*	115

Γ' is the minimum ratio of target mass to projectile mass causing catastrophic disruption at 10 km/s. Γ is the ratio of ejected mass to projectile mass in a noncatastrophic collision at 10 km/s.
*No tests have been performed to obtain this value. This lower limit follows from the definitions of Γ' and Γ.

where K and η are constants. From tests performed on basalt, $K = 0.4$, and $\eta = -0.8$ [Gault et al., 1963; Dohnanyi, 1971]. From these results and associated modeling [Kessler, 1971] it was concluded that the asteroid belt must include particles as small as dust grains. Of course, the objects in the asteroid belt are solid chunks of material, unlike the anthropogenic satellites in earth orbit.

IMPACTS INTO SPACECRAFT STRUCTURES

The objects expected to collide in earth orbit consist of predominantly scientific and military satellites, rocket motors, and fragments of the same caused by malfunctions or deliberate destruction. The proportion of solid chunks of material will be very small, and the majority of collisions will be between open and closed structures filled with equipment. Thus the typical object will be a nonhomogeneous mass with discontinuities and many voids. The only reported hypervelocity tests where fragment distributions were obtained for 'typical' space structures, with internal components, were performed by Langley Research Center [Bess, 1975]. The tests showed that for these particular configurations the same general ejected mass and fragment size distribution laws established for the solid objects also apply to spacecraft structures.

When these results were scaled to 10 km/s, a value of $\Gamma = 115$ was obtained for spacecraft structures. That is, in a noncatastrophic collision between a spacecraft structure and smaller object at 10 km/s the ejected mass would be 115 times the mass of the smaller object. Note from Table 1 that this value is not too different from that for solid 1100-0 aluminum. No tests have been performed to try to duplicate a catastrophic collision involving a spacecraft structure. Obviously, if the crater produced in a 'noncatastrophic' collision is larger than the satellite, then the collision is catastrophic; thus Γ' must be greater than 115 and by analogy to solid objects is likely to be much larger than 115.

When normalized to the total ejected mass, as in (7), the distribution of fragments from spacecraft structures looks very similar to that of basalt. Values of $K = 0.89$ and $\eta = -0.8$ were obtained from one test into the spacecraft structure, and $K = 0.69$ and $\eta = -0.84$ were obtained from the other test [Bess, 1975]. The velocities of fragments, measured from a 400-frame/s film, were found to be very slow, about 10–30 m/s. Most of the fragment mass from basalt targets is slower than 100 m/s [Zook, 1967].

Thus for collisions between earth-orbiting objects the following relationship was adopted:

$$M_e = 115 M_1 \qquad (8)$$

when $M_2 > 115 M_1$. If $M_2 < 115 M_1$, then $M_e = M_2$ (the mass of M_1 was assumed to be small or lost). The number of fragments of mass M and larger resulting from the collision is given by

$$N = 0.8(M/M_e)^{-0.8} \qquad (9)$$

SATELLITE MASS

Before these impact equations can be used, the individual mass of each satellite is required. Some of these masses are available, but not in one source. Mass and areas for a few payloads and rocket motors were found in several publications [Bowman, 1963; Corliss, 1967; Martin, 1967; Von Braun and Ordway, 1975; Aerospace Defense Command, 1977], and these data are plotted in Figure 5. Also plotted is the debris resulting from the breakup of a Centaur D-IT rocket [Drago and Edge-

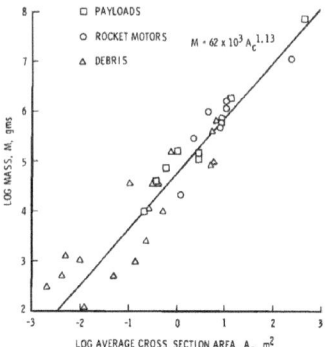

Fig. 5.   Satellite mass versus average cross section.

combe, 1974]. The data were found to fit the relationship

$$M = 62 \times 10^3 A_c^{1.13} \qquad (10)$$

where M is the mass of the satellite in grams and A_c is its average cross section in square meters. The slope of this fit was expected to be between 1.0 (which would be true of hollow structures with constant thickness walls) and 1.5 (which would be true of structures having a constant mass density). The value of 1.13 fell within this range.

Therefore masses were assigned to each of the sizes shown in Figure 3, according to (10). The amount of mass ejected in collisions between each size was calculated by using (8); however, the ejected mass was not allowed to exceed the mass of the larger object (the target mass). Each collision was then weighted according to its probability of occurrence in order to obtain an average mass ejected per collision. A result of 870 kg/collision was obtained. A detailed examination revealed that most mass resulted from satellites between 16 and 40 m² being impacted by satellites of 0.25 m² and larger; thus the ejected mass was limited in many cases by the mass available in the larger object.

This suggested that the average ejected mass may be fairly insensitive to (8). This sensitivity was tested by allowing the constant to vary by a factor of 10 on either side of 115, resulting in the average ejected mass varying from 440 to 1190 kg/collision. A more realistic lower limit for the constant in (8) is about a factor of 4 less (resulting in an average of 600 kg/ collision), while the upper limit may be a factor of 10 or more larger, owing to the concept of catastrophic collisions (i.e., the value of Γ' in (4) could be 1150 or larger). Thus the measured size distribution of debris causes the average ejected mass to be fairly insensitive to the uncertainty in mass ejected during a collision. The sensitivity would increase if a sufficient number of objects smaller than 0.25 m² were known to be in earth orbit.

Thus on the basis of currently observed distribution of satellites an average of $M_e = 8.7 \times 10^8$ g is ejected in each collision shown in Figure 4.

AVERAGE DEBRIS FLUX BETWEEN 700 AND 1200 KM

By integrating (2) over the region of space between 700 and 1200 km a current collision rate of 0.01/yr was obtained, compared to 0.013/yr for all of space. Thus about 77% of the

collisions were found to occur within this volume. The near-circular orbit of most objects found within this volume combined with the low ejection velocity of most fragments was justification for assuming that 77% of the collisional fragments would also be found within this volume.

Thus an average debris flux for this volume of space was found by

$$F = (N/U)\bar{V}_s \qquad (11)$$

where N is the average number of objects of mass M and larger found within volume U and F is the flux of debris of mass M and larger. This flux was computed as a function of time by increasing the 3866 satellites (1976 number) at the 'nominal' rate of 510/yr. It was calculated that 51% of these satellites are normally found within the volume of space between 700 and 1200 km. The collision rate within this volume was assumed to be 77% of that shown in Figure 4. The number of fragments generated by each collision is given by (9) where $M_e = 8.70 \times 10^8$.

The projected debris flux is shown in Figure 6 for the years 1990, 2020, and 2100, compared with the natural meteoroid flux [*Cour-Palais*, 1969] and the current (1976) debris flux. The curve for the current debris flux is flat for debris masses less than 200 g only because these sizes cannot currently be observed. Even so, for meteoroid mass greater than 0.3 g the current debris flux between 700 and 1200 km already exceeds the natural meteoroid flux. As is illustrated in Figure 6, a significant number of debris fragments smaller than 200 g will be generated in the future, further exceeding the meteoroid flux.

To illustrate the effect on future space missions, consider three types of missions. The first is an unmanned satellite having an average cross section of 10 m² (i.e., about 3.6-m diameter) and a desired average lifetime of 10 years. The area-time product of such a satellite would be 10² m² yr, so that the design flux, to ensure an average 10-yr lifetime, would be 10^{-2}/ m² yr. If meteoroids are the only hazard, then Figure 6 predicts that the satellite should be designed to survive a 2×10^{-4} g

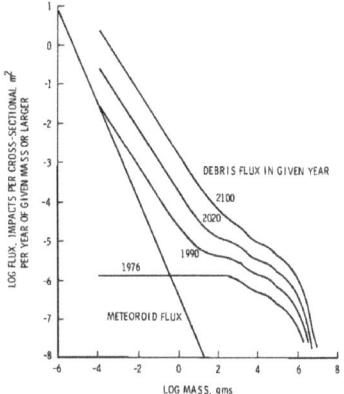

Fig. 6.   Average debris flux between 700 and 1200 km; assumes that (1) the net satellite input rate is always 510/yr and (2) there is no atmospheric drag.

TABLE 2.   Design Requirements for Three Hypothetical Missions

Mission Type	Unmanned	Skylab	Space Station
Design flux, impacts/m² yr	10⁻³	10⁻⁴	10⁻⁶
Design meteoroid impact mass, g	2×10^{-4}	1×10^{-2}	4×10^{-1}
Year 2020, 1200-km altitude, debris design impact mass, g	1×10^{-2}	2.0	5×10^{6}
Altitude band where equilibrium debris flux exceeds meteoroid flux (assuming change to zero net satellite input after given year), km			
1980	none	750–1200	500–1200
2020	800–1200	550–1200	400–1500

meteoroid impact. The second type is a manned spacecraft having an average cross section of 100 m², a mission duration of 1 yr, and a desired probability of impact damage of less than 0.01. In this case the design flux would be 10^{-4}/m² yr, or about the same as the 1973–1974 Skylab mission. A meteoroid shield, weighing over 300 kg, was added to the Skylab structure in order to protect it against 10^{-2} g meteoroid impacts. The third type of mission is a large space station having an average cross section of 10,000 m² (i.e., a little over 100-m diameter), a mission duration of 10 yr, and a desired probability of impact damage of less than 0.1. The design flux would be 10^{-6}/m² yr, requiring protection against a 0.4-g meteoroid. These conditions are summarized in Table 2.

However, under certain conditions, as illustrated in Figure 6, the debris flux for these missions may exceed the meteoroid flux. By the year 2020 the unmanned, Skylab, and space station type missions may require protection against a 1×10^{-2} g, 2 g, and 5×10^{6} g debris particle, respectively. These missions would require more weight for impact protection. However, protection requirements against even a 100-g impact are so severe that a space station may have to either accept a much higher probability of impact damage or be restricted to altitudes where the debris flux is lower.

The increased risk of impact damage may lead to certain constraints being placed on launched satellites in order to reduce the projected debris hazard. For purposes of illustration it was assumed that beginning in 2020, the net satellite input rate of 510/yr is changed to zero. A zero rate can be maintained by ceasing all launch activity, returning a similar object for every object placed into orbit, or causing the reentry of unused objects. The results of this assumption are shown in Figure 7. Notice that the flux of fragments continues to increase after the year 2020.

OTHER SOURCES AND SINKS FOR DEBRIS

The effects of catastrophic collisions, collisions involving fragments from previous collisions, and the current number of unobserved small fragments all represent other sources of debris, increasing the results presented thus far. Each source has been looked at in some detail, and each is found to have a relatively small effect. (1) As was previously stated, the amount of mass produced per collision is mostly limited by the amount of mass available; thus the concept of catastrophic collisions could only about double the amount of fragment mass per collision. (2) With time, enough collisional fragments could be produced to become important in producing new collisional fragments. When these conditions apply, the number of objects will increase exponentially with time, even though no new objects may be placed into orbit by man. Some

preliminary analysis indicated that the uncertainty in timing of this phenomenon is large but probably of the order of several hundred years. (3) The presence of objects that are too small to be detected by ground radar would imply that the current debris flux should be increased correspondingly. These objects must already exist as the results of numerous satellite explosions and other types of debris [Neste et al., 1976]. However, a preliminary analysis concluded that the current number would have to be higher than the 1990 or 2020 projected number before the objects would become significant contributors to the collision-fragmentation process. Thus while the 18% deficiency observed in the SDC catalog [Hendren and Anderson, 1976] means that the current debris flux should be increased by 18%, the projected debris resulting from fragmentation would be increased by much less that 18%. In fact, if the unobserved population of small fragments requires that the total number of observed satellites be increased by a factor of 2.5, as suggested by Brooks et al. [1975], then the number of collisional fragments would be increased by less than 3%.

While these three additional sources may represent a small effect when they are taken individually, they may combine to produce a significant effect. For example, a 2.5 factor increase in the observed population, combined with a more realistic

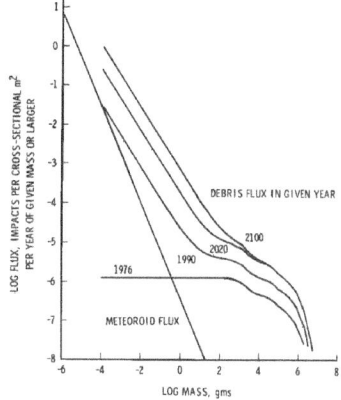

Fig. 7.   Average debris flux between 700 and 1200 km; assumes that (1) the net satellite input rate changes from 510/yr to zero in the year 2020 and (2) there is no atmospheric drag.

180 Sky Alert

value of catastrophic disruption mass ratio Γ', could double the number of collision fragments. Also, the exponential increase in fragments with time (source 2) could be observed much earlier than several hundred years, depending on the nature of catastrophic collisions, the projected number and size of future satellites, and the current number of unobserved small fragments. Meaningful analysis of this type must await more data.

Only a few 'sinks,' or removal mechanisms, exist for earth-orbiting satellites. They are basically only retrieval and atmospheric reentry. One could argue that as a result of catastrophic collisions, large objects disappear, and thus collision is a sink for these objects. However, hundreds of years are required before this becomes a significant sink, whereas collision is a much more important source of small fragments at a much earlier date. Thus for the near future it is accurate to think of collisions as only a source.

If retrieval is implemented, it could significantly alter the conclusions reached thus far. Collision rates are proportional to the collision cross section of satellites. Figure 3 reveals that 90% of the satellite area is contained in 20% of the satellites. Thus removal of large satellites could effectively slow down the collision rate. However, as the number of collision fragments increases, the concentration of area will move toward the more numerous, smaller objects, making it more difficult to slow down the fragmentation process by retrieval.

Atmospheric drag will eventually cause the reentry of many objects in orbit. Drag acts most quickly on small low-altitude objects and is a significant factor in reducing the number of satellites below 400 km, as is shown in Figure 1. The projected number of small fragments between 700 and 1200 km will be reduced from that shown in Figures 6 and 7 by the effects of atmospheric drag. However, since the collision frequency is low at altitudes less than 700 km, atmospheric drag will act as the primary source of fragments at these lower altitudes. That is, drag will act to remove collision fragments from the 700- to 1200-km region and drag them through these lower altitudes. Thus since atmospheric drag is the only natural sink for debris and since it may be an important source of debris at lower altitudes, it deserves a more detailed analysis.

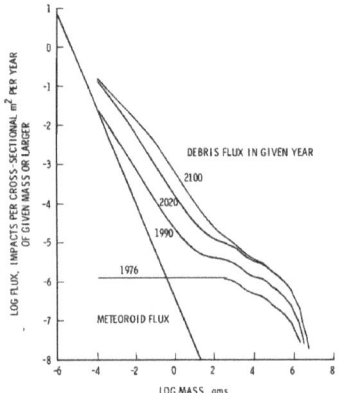

Fig. 8. Average debris flux at 1200 km; assumes that the net satellite input rate changes from 510/yr to zero in the year 2020.

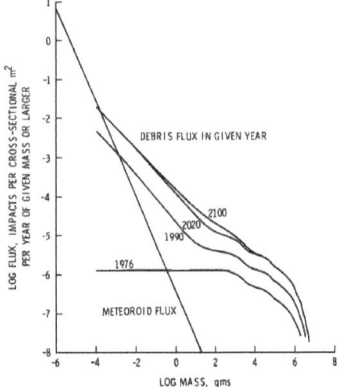

Fig. 9. Average debris flux at 800 km; assumes that the net satellite input rate changes from 510/yr to zero in the year 2020.

ATMOSPHERIC DRAG

Atmospheric drag will act to first circularize an orbit, then cause the object to spiral into the atmosphere. The speed at which this process works is proportional to the area to mass ratio of the object, as well as the atmospheric density at a given altitude [*Martin*, 1967]. With the use of the atmospheric model of *Lou* [1973], energy loss rates were calculated to obtain orbital decay times. The time for a 1-cm-radius sphere of mass density 2 g/cm³ (mass of 8.4 g) to change its altitude by 100 km for circular orbits of 800 km and 1200 km was calculated to be 32 yr and 455 yr, respectively. This compares to 110 yr and 2000 yr, respectively, calculated by *Martin* [1967] and 20 yr and 100 yr, respectively, calculated by *Brooks et al.* [1975]. The large range in values results from uncertainties in the atmospheric model and drag coefficients.

Lifetimes at a particular altitude were assumed to be the same as the calculated values for a circular orbit to decrease by 100 km. Of course, actual lifetimes could be longer or shorter, depending on orbital perigee or apogee. However, since most of the satellites within the 700- to 800-km band are in nearly circular orbits, this approximation should be fairly accurate. Lifetime as a function of particle size varies as particle radius, assuming a constant mass density.

Thus a model was developed which calculated the number of collisional fragments of a particular size which are produced during the lifetime of that size. Figure 8 gives the results of that model at 1200-km altitude. By comparison to Figure 7, drag had no effect on the debris design impact masses given in Table 1 for the year 2020. The change to a zero net satellite input rate in 2020 still led to an increased debris flux in 2100, although the flux of debris particles of less than 0.1 g was reduced by atmospheric drag.

Figure 9 gives the results of atmospheric drag at 800 km. In comparison with Figure 7 the debris design impact mass for an unmanned mission in the year 2020 was reduced, although it was still higher than the meteoroid design impact mass. For Skylab and space station type missions the debris flux was essentially unchanged by atmospheric drag. Of particular interest in Figure 9 is that a change in the net satellite input rate to zero in the year 2020 resulted in a near-equilibrium being established for times after that date. That is, collisional fragments are being generated at the same rate as they are being removed by atmospheric drag. This equilibrium was reached almost immediately after the year 2020 for sizes smaller than about 0.1 g, while the sizes between 0.1 and 10^4 g were near equilibrium in 2020, reaching it by the year 2100. By comparing Figures 8 and 9 an equilibrium at 1200 km was reached for sizes smaller than 0.1 g by the year 2100 but at a flux level about a factor of 10 larger than that at 800 km.

These equilibrium debris levels suggest another way of looking at the future debris fluxes: that is, given a date to change from a net satellite input rate of 510/yr to zero, what will the equilibrium debris flux eventually become? For altitudes below 800 km the equilibrium will be reached almost immediately after the change to zero input date, whereas for altitudes up to 1200 km, several hundred years may be required, especially for the larger debris sizes. Figure 10 shows the average equilibrium debris flux at 800 km for various change input rate dates. Notice that even an early (1980) change to zero net input rate would eventually affect Skylab type missions at 800 km, whereas unmanned missions would not be affected at this altitude until a change after 2020. A similar curve for 1200-km altitude would be about a factor of 10 higher. The time required for atmospheric drag to drag a fragment through a particular altitude is inversely proportional to the atmospheric density at that altitude. Thus the average equilibrium debris flux at altitudes below 800 km (and actually above 800 km, up to 1200 km), was found by taking the ratio of the atmospheric density at that altitude to the atmospheric density at 800 km. Figure 11 gives that ratio of equilibrium fluxes. Thus for example, the equilibrium flux at 500 km was found by reducing the debris fluxes in Figure 10 by a factor of 30.

From this type of analysis the region of space where the debris flux could exceed the natural meteoroid flux was determined for the three types of missions. Table 2 summarizes these altitude bands, assuming a change to the zero net input rate in 1980 or 2020. An early (1980) change would prevent

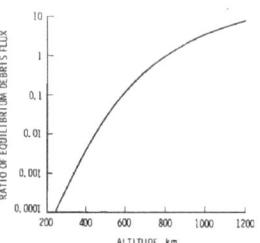

Fig. 11.   Ratio of equilibrium debris flux at given altitude to flux at 800 km.

unmanned satellites from being affected by the debris flux anywhere in space. Skylab and space station type missions would be affected between 750 and 1200 km and 500 and 1200 km, respectively. However, if the change does not occur until the year 2020, the region of space where unmanned missions would be affected becomes 800–1200 km, and the Skylab and space stations regions affected expand to 550–1200 km and 400–1500 km, respectively. Thus an increase in the change date leads to larger regions of space where the debris flux could eventually exceed the meteoroid flux. The only obvious method of lowering this eventual equilibrium flux is to minimize the number of large satellites, either by a change in launch practices or by retrieval.

THE 'POTENTIAL' DEBRIS ENVIRONMENT

Thus far, the physical processes of collision and atmospheric drag have been applied to the observed distribution of satellites in earth orbit in an attempt to predict the time history of debris in earth orbit. However, it is known that satellites can fragment from other sources and that attempted modeling of incomplete data may predict the future debris flux inadequately.

Another approach might be to define the 'potential' debris flux and then develop arguments which would prevent this potential from being realized. This type of analysis would produce flux curves much higher than the previous curves, but the analysis would predict accurately what is possible, although not necessarily what is probable. A beginning point would be to assume that all satellites, through some unknown mechanism, become fragmented into some preferred but yet to be determined fragment size. The potential flux as a function of mass is then found by assuming that all of these fragments are of the preferred size. This potential flux is actually an envelope of single-size fluxes. As data become available that indicate that only some fraction of the total mass goes into a particular size interval, then the potential flux could be lowered appropriately in that size interval.

The average satellite mass was found (by using Figure 3 and (10)) to be 1.3×10^6 g, or a total 1976 satellite mass of 5×10^9 g. If this total mass were to fragment into debris of mass M, then the number of fragments would be

$$N = 5 \times 10^9 M^{-1} \qquad (12)$$

If these fragments maintained the same orbits as the original 3866 satellites, then by using the same techniques as the original model, the 'average potential flux' between 700 and 1200 km was found and is shown in Figure 12. Note that if this

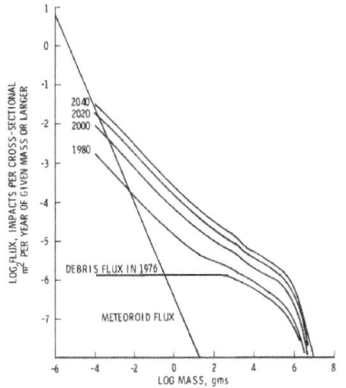

Fig. 10.   Average equilibrium debris flux at 800 km; assumes that the net satellite input rate changes from 510/yr to zero in given year.

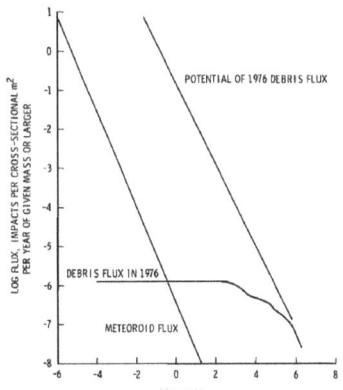

Fig. 12.   Potential debris flux between 700 and 1200 km. Potential flux assumes that all satellites are fragmented into given mass.

potential were ever realized, all types of missions into this region would experience a flux level far in excess of the natural meteoroid flux. In fact, since it becomes impractical to protect against impacts larger than about 100 g, all missions would have to expect damage in certain regions of space.

The potential flux for other distances may be scaled from Figure 1. For example, the spatial density at 300 km was found from Figure 1 to be about a factor of 30 lower than the average between 700 and 1200 km; thus the debris curves in Figure 12 could be lower by this amount to obtain the 300-km fluxes.

As more mass is added to the system with time, the potential flux will increase correspondingly. The potential flux can be decreased as data become available. For example, at least 31 larger satellites, or about 1% of the total number of satellites, have already exploded in orbit [Neste et al., 1976]. If fragmentation could be limited to these 31 explosions, then the potential flux could be reduced by a factor of 100. In addition, if it were found that only a small fraction of the fragmented mass were of a particular size or larger, then the potential flux for that size could be reduced further.

However, if fragmentation is maintained at 1%, or about 5 explosions per year, the potential flux would continue to increase until an equilibrium is reached with atmospheric drag. At 800 km this equilibrium would be about 4% of the potential flux shown in Figure 12 for fragments of 8.4 g and 0.4% for fragments of 8.4×10^{-8} g. Thus without meaningful data on the actual size distribution of fragments, the expected fragmentation rate, and the lifetime of these fragments it is difficult to lower the potential debris flux significantly.

FUTURE ANALYSIS REQUIRED

In order to reduce the uncertainty in the projected debris environment, additional data will be required on the effects of hypervelocity collisions between spacecraft, as well as the effects of other spacecraft fragmentation processes. Additional data on the number of small objects in space can be obtained by a detailed examination of individual launch and orbit injection procedures. The results of these studies should be tested by experiments designed to detect objects in orbit smaller than

10 cm. An optical experiment is described by Neste et al. [1976] which would detect debris in the 1-mm to 10-cm range. With the availability of improved input data, a model could be developed to include the results from explosions, the effects of collisions resulting from collision fragments, the effects of catastrophic collisions, and orbital changes resulting from collisions.

Various methods to stop or slow the formation of a debris belt should be studied. The model suggests that the most effective way would be to keep the number of large objects as small as practical. This could be accomplished by planning launches so that large objects can be caused to reenter when their usefulness is complete or by using the space shuttle concept to retrieve objects in orbit which no longer serve a useful function. Since it is impractical to retrieve the much larger number of large and small fragments, every effort should be made to prevent their production in space, either by explosion or by collision.

The evolution of the debris belt should be followed to its conclusion. As was pointed out by Alfvén and Arrhenius [1976], the consequence of many collisions is to change the orbits of objects to be more alike. This process may be responsible for creating 'jet streams' in the asteroid belt. With time, according to Alfvén, these jet stream orbits will become identical, and a single planet will accrete (billions of years in the future) at 2.8 AU. However, in the case of the earth the debris belt will be within the Roche's limit (less than 9000-km altitude), preventing accretion into a single object. Thus the end result, assuming that drag does not act fast enough, could be a ring system, similar to that around Saturn and Uranus.

CONCLUSIONS

A model has been developed which considers the major source and sink terms for the growth of the satellite population in earth orbit. While significant uncertainties exist, the following conclusions, if current trends continue, seem unavoidable:

1. Collisional breakup of satellites will become a new source for additional satellite debris in the near future, possibly well before the year 2000.

2. Once collisional breakup begins, the debris flux in certain regions near earth may quickly exceed the natural meteoroid flux.

3. Over a longer time period the debris flux will increase exponentially with time, even though a zero net input rate may be maintained.

4. The processes which will produce these fragments are totally analogous to the processes that probably occurred in the formation of the asteroid belt but require a much shorter time.

Effective methods exist to alter the current trend without significantly altering the number of operational satellites in orbit. These methods include reducing the projected number of large, nonoperational satellites and improved engineering designs which reduce the frequency of satellite breakups from structural failure and explosions in space. Delay in implementation of these methods reduces their effectiveness.

Acknowledgments.   We thank Joe M. Alvarez (NASA) for helpful discussions of past work in these areas, leading to further helpful discussion with Preston M. Landry (NORAD) and Larry Rice (SAI). We acknowledge suggestions by Herbert A. Zook and Andrew E. Potter (NASA) which led to changes in the content of this paper.
The Editor thanks D. R. Brooks and V. J. Drago for their assistance in evaluating this paper.

2646 KESSLER AND COUR-PALAIS: ARTIFICIAL SATELLITE DEBRIS BELT

REFERENCES

Aerospace Defense Command, Average radar cross section (sq. m.) as of 1 July 1977, report, Peterson Air Force Base, Colo., 1977.

Alfvén, H., and G. Arrhenius, Evolution of the solar system, *NASA Spec. Publ. SP-345*, 1976.

Bess, T. D., Mass distribution of orbiting man-made space debris, *NASA Tech. Note TND-8108*, 1975.

Bowman, N. J., *The Handbook of Rockets and Guided Missiles*, Perastadion Press, Newtown Square, Pa., 1963.

Brooks, D. R., G. G. Gibson, and T. D. Bess, Predicting the probability that earth-orbiting spacecraft will collide with man-made objects in space, in *Space Rescue and Safety*, pp. 79–139, American Astronautical Society Publications Office, Tarzana, Calif., 1975.

Corliss, W. R., Scientific satellites, *NASA Spec. Publ. SP-133*, 1967.

Cour-Palais, B. G., Meteoroid environment model—1969 (near earth to lunar surface), *NASA Spec. Publ. SP-8013*, 1969.

Dohnanyi, J. S., Fragmentation and distribution of asteroids, Physical Studies of Minor Planets, *NASA Spec. Publ. SP-267*, 263–295, 1971.

Drago, V. J., and D. S. Edgecombe, A review of NASA orbital decay reentry debris hazard, *Rep. BMI-NVLP-TM-74-1*, Battelle Mem. Inst., Columbus, Ohio, 1974.

Gault, D. E., E. M. Shoemaker, and H. J. Moore, Spray ejected from the lunar surface by meteoroid impact, *NASA Tech. Note TND-1767*, 1963.

Hendren, J. K., and A. Anderson, Comparison of the perimeter acquisition radar (PAR) satellite track capability to the Space Defense Center (SDC) satellite catalogue—Unknown satellite track experiment, *Rep. SAI-77-701-HU*, Sci. Appl., Inc., Huntsville, Ala., 1976.

Kessler, D. J., Estimate of particle densities and collision danger for spacecraft moving through the asteroid belt, Physical Studies of Minor Planets, *NASA Spec. Publ. SP-267*, 595–605, 1971.

Lou, G. Y., Models of earth's atmosphere (90 to 2500 km), *NASA Spec. Publ. SP-8021*, 1973.

Martin, C. N., *Satellites Into Orbit*, translated from French by T. Schoeters, George G. Harrap Co. Ltd., London, 1967.

McCarter, J. W., Probability of satellite collision, *NASA Tech. Memo TMX-64671*, 1972.

Moore, H. J., D. E. Gault, and E. D. Heitowit, Change of effective target strength with increasing size of hypervelocity impact craters, Proceedings of 7th Hypervelocity Impact Symposium, vol. IV, Theory, p. 35, U.S. Army, Navy, Air Force, 1965. (Available from National Technical Information Service, Springfield, Va.)

NASA, Satellite Situation Report, vol. 14, Off. of Publ. Aff., Goddard Space Flight Center, Greenbelt, Md., 1974.

NASA, Satellite Situation Report, vol. 15, Off. of Publ. Aff., Goddard Space Flight Center, Greenbelt, Md., 1975.

NASA, Satellite Situation Report, vol. 16, Off. of Publ. Aff., Goddard Space Flight Center, Greenbelt, Md., 1976.

Neste, S. L., R. K. Soberman, K. Lichtenfield, and L. R. Eaton, The Sisyphus system—Evaluation, suggested improvements, and application to measurements of space debris, final report, contract NAS1-13407, Gen. Elec. Co., Philadelphia, Pa., 1976.

Ruck, G. T., D. E. Barrick, W. D. Stuart, and C. K. Krichbaum, *Radar Cross Section Handbook*, Plenum, New York, 1970.

Von Braun, W., and F. I. Ordway III, *History of Rockets and Space Travel*, Thomas Y. Crowell Company, New York, 1975.

Zook, H. A., The problem of secondary ejecta near the lunar surface, in *Saturn V/Apollo and Beyond*, pp. EN-8-1 to EN-8-24, American Astronautical Society Publications Office, Tarzana, Calif., 1967.

(Received November 7, 1977;
revised February 10, 1978;
accepted February 22, 1978.)

C (Selected) Spacecraft Failures and Anomalies Due to Solar and Geomagnetic (Solar Event-Induced) Storms or Orbital Debris Impacts

Intelsat K (1994)
A geomagnetic storm caused the satellite to wobble and produce fluctuations in its antenna coverage. The satellite recovered after activating a backup system.

Anik E-1 (1994)
Telsat Canada communications satellite began to spin out of control due to the same storm that caused the problems with Intelsat K (above). Backup systems were brought online and the satellite recovered in 8 hr.

Anik E-2 (1994)
Telsat Canada communications satellite began to spin out of control due to the same geomagnetic storm that caused the problems with Intelsat K. The backup systems failed to restore the satellite's operation. Several months later, operation of the satellite resumed after a work-around was devised by ground system engineers.

Maritime European Communications Satellite (1982)
The satellite's pointing system went into "safe" mode, shutting down all communication systems. The failure was due to electrostatic discharges associated with a geomagnetic storm occurring at the time.

The US Air Force Defense Space Communications Satellite 9431 (1973)
Power to the satellite's communications system was interrupted. The failure was due to a high-energy electrical discharge caused by a geomagnetic storm.

Small Expendable Deployer System Second Flight (1994)
This 20-km tether experiment ended abruptly when the tether was severed by an orbital debris impact.

Miniature Sensor Technology Integration Satellite (1994)
Contact with the satellite was lost after it was impacted by a piece of orbital debris.
Space Shuttle Flight STS-45 (1992)
The Space Shuttle *Atlantis* experienced two impacts on the leading edge of its right wing.

Space Shuttle Flight STS-49 (1992)
The window was chipped by impact with a small piece of orbital debris.

KOSMOS-1275 (1981)
The KOSMOS-1275 broke up into over 200 trackable fragments as a result of being hit by a piece of orbital debris.

National Oceanic and Atmospheric Administration NOAA-9 (1984)
High solar activity caused problems with the spacecraft's attitude control system.

Reference

Beddingfield, K.L.; Leach, R.D.; Alexander, M.B. Spacecraft System Failures and Anomalies Attributed to the Natural Space Environment. NASA Reference Publication 1390, August 1996.

D International Agreements Governing Orbital Debris and Space Weaponization

Legal Implications for Orbital Debris Removal

There are treaties in place that govern the use of outer space and most countries of the world have signed them. The most significant, the "Treaty on Principles Governing the Activities of States in the Exploration and Use of Outer Space, including the Moon and Other Celestial Bodies", or, more commonly, "The Outer Space Treaty", was enacted by the General Assembly of the United Nations in 1967. In it are the general principles which allow satellites to fly over any land on the surface of the Earth unimpeded – a right not generally understood or agreed to until this treaty was enacted. (Some nations originally claimed that they owned the airspace and outer space that extended above their national boundaries into the infinite depths of space!) It is this treaty that is often cited in discussions of orbital debris removal and space weaponization and the text of it is reproduced below [1].

> "Treaty on Principles Governing the Activities of States in the Exploration and Use of Outer Space, including the Moon and Other Celestial Bodies
>
> The States Parties to this Treaty,
>
> Inspired by the great prospects opening up before mankind as a result of man's entry into outer space,
>
> Recognizing the common interest of all mankind in the progress of the exploration and use of outer space for peaceful purposes,
>
> Believing that the exploration and use of outer space should be carried on for the benefit of all peoples irrespective of the degree of their economic or scientific development,
>
> Desiring to contribute to broad international cooperation in the scientific as well as the legal aspects of the exploration and use of outer space for peaceful purposes,
>
> Believing that such cooperation will contribute to the development of mutual understanding and to the strengthening of friendly relations between States and peoples,

Recalling resolution 1962 (XVIII), entitled "Declaration of Legal Principles Governing the Activities of States in the Exploration and Use of Outer Space", which was adopted unanimously by the United Nations General Assembly on 13 December 1963,

Recalling resolution 1884 (XVIII), calling upon States to refrain from placing in orbit around the Earth any objects carrying nuclear weapons or any other kinds of weapons of mass destruction or from installing such weapons on celestial bodies, which was adopted unanimously by the United Nations General Assembly on 17 October 1963,

Taking account of United Nations General Assembly resolution 110 (II) of 3 November 1947, which condemned propaganda designed or likely to provoke or encourage any threat to the peace, breach of the peace or act of aggression, and considering that the aforementioned resolution is applicable to outer space,

Convinced that a Treaty on Principles Governing the Activities of States in the Exploration and Use of Outer Space, including the Moon and Other Celestial Bodies, will further the purposes and principles of the Charter of the United Nations,

Have agreed on the following:

Article I
The exploration and use of outer space, including the Moon and other celestial bodies, shall be carried out for the benefit and in the interests of all countries, irrespective of their degree of economic or scientific development, and shall be the province of all mankind.

Outer space, including the Moon and other celestial bodies, shall be free for exploration and use by all States without discrimination of any kind, on a basis of equality and in accordance with international law, and there shall be free access to all areas of celestial bodies.

There shall be freedom of scientific investigation in outer space, including the Moon and other celestial bodies, and States shall facilitate and encourage international cooperation in such investigation.

Article II
Outer space, including the Moon and other celestial bodies, is not subject to national appropriation by claim of sovereignty, by means of use or occupation, or by any other means.

Article III
States Parties to the Treaty shall carry on activities in the exploration and use of outer space, including the Moon and other celestial bodies,

in accordance with international law, including the Charter of the United Nations, in the interest of maintaining international peace and security and promoting international cooperation and understanding.

Article IV
States Parties to the Treaty undertake not to place in orbit around the Earth any objects carrying nuclear weapons or any other kinds of weapons of mass destruction, install such weapons on celestial bodies, or station such weapons in outer space in any other manner.

The Moon and other celestial bodies shall be used by all States Parties to the Treaty exclusively for peaceful purposes. The establishment of military bases, installations and fortifications, the testing of any type of weapons and the conduct of military maneuvers on celestial bodies shall be forbidden. The use of military personnel for scientific research or for any other peaceful purposes shall not be prohibited. The use of any equipment or facility necessary for peaceful exploration of the Moon and other celestial bodies shall also not be prohibited.

Article V
States Parties to the Treaty shall regard astronauts as envoys of mankind in outer space and shall render to them all possible assistance in the event of accident, distress, or emergency landing on the territory of another State Party or on the high seas. When astronauts make such a landing, they shall be safely and promptly returned to the State of registry of their space vehicle.

In carrying on activities in outer space and on celestial bodies, the astronauts of one State Party shall render all possible assistance to the astronauts of other States Parties.

States Parties to the Treaty shall immediately inform the other States Parties to the Treaty or the Secretary-General of the United Nations of any phenomena they discover in outer space, including the Moon and other celestial bodies, which could constitute a danger to the life or health of astronauts.

Article VI
States Parties to the Treaty shall bear international responsibility for national activities in outer space, including the Moon and other celestial bodies, whether such activities are carried on by governmental agencies or by non-governmental entities, and for assuring that national activities are carried out in conformity with the provisions set forth in the present Treaty. The activities of non-governmental entities in outer space, including the Moon and other celestial bodies, shall require authorization and continuing supervision by the appropriate State Party to the Treaty. When activities are

carried on in outer space, including the Moon and other celestial bodies, by an international organization, responsibility for compliance with this Treaty shall be borne both by the international organization and by the States Parties to the Treaty participating in such organization.

Article VII
Each State Party to the Treaty that launches or procures the launching of an object into outer space, including the Moon and other celestial bodies, and each State Party from whose territory or facility an object is launched, is internationally liable for damage to another State Party to the Treaty or to its natural or juridical persons by such object or its component parts on the Earth, in air space or in outer space, including the Moon and other celestial bodies.

Article VIII
A State Party to the Treaty on whose registry an object launched into outer space is carried shall retain jurisdiction and control over such object, and over any personnel thereof, while in outer space or on a celestial body. Ownership of objects launched into outer space, including objects landed or constructed on a celestial body, and of their component parts, is not affected by their presence in outer space or on a celestial body or by their return to the Earth. Such objects or component parts found beyond the limits of the State Party to the Treaty on whose registry they are carried shall be returned to that State Party, which shall, upon request, furnish identifying data prior to their return.

Article IX
In the exploration and use of outer space, including the Moon and other celestial bodies, States Parties to the Treaty shall be guided by the principle of cooperation and mutual assistance and shall conduct all their activities in outer space, including the Moon and other celestial bodies, with due regard to the corresponding interests of all other States Parties to the Treaty. States Parties to the Treaty shall pursue studies of outer space, including the Moon and other celestial bodies, and conduct exploration of them so as to avoid their harmful contamination and also adverse changes in the environment of the Earth resulting from the introduction of extraterrestrial matter and, where necessary, shall adopt appropriate measures for this purpose.

If a State Party to the Treaty has reason to believe that an activity or experiment planned by it or its nationals in outer space, including the Moon and other celestial bodies, would cause potentially harmful interference with activities of other States Parties in the peaceful exploration and use of outer space, including the Moon and other celestial bodies, it shall undertake appropriate international consultations before proceeding with any such activity or experiment. A State Party to the Treaty which has reason to believe that an activity or experiment planned by another State Party in outer space, including the Moon and other celestial bodies, would cause potentially harmful interference with activities in the peaceful exploration and use of outer space, including the Moon and other celestial bodies, may request consultation concerning the activity or experiment.

Article X
In order to promote international cooperation in the exploration and use of outer space, including the Moon and other celestial bodies, in conformity with the purposes of this Treaty, the States Parties to the Treaty shall consider on a basis of equality any requests by other States Parties to the Treaty to be afforded an opportunity to observe the flight of space objects launched by those States. The nature of such an opportunity for observation and the conditions under which it could be afforded shall be determined by agreement between the States concerned.

Article XI
In order to promote international cooperation in the peaceful exploration and use of outer space, States Parties to the Treaty conducting activities in outer space, including the Moon and other celestial bodies, agree to inform the Secretary-General of the United Nations as well as the public and the international scientific community, to the greatest extent feasible and practicable, of the nature, conduct, locations and results of such activities. On receiving the said information, the Secretary-General of the United Nations should be prepared to disseminate it immediately and effectively.

Article XII
All stations, installations, equipment and space vehicles on the Moon and other celestial bodies shall be open to representatives of other States Parties to the Treaty on a basis of reciprocity. Such representatives shall give reasonable advance notice of a projected visit, in order that appropriate consultations may be held and that maximum precautions may be taken to assure safety and to avoid interference with normal operations in the facility to be visited.

Article XIII
The provisions of this Treaty shall apply to the activities of States Parties to the Treaty in the exploration and use of outer space, including the Moon and other celestial bodies, whether such activities are carried on by a single State Party to the Treaty or jointly with other States, including cases where they are carried on within the framework of international intergovernmental organizations. Any practical questions arising in connection with activities carried on by interna-

tional intergovernmental organizations in the exploration and use of outer space, including the Moon and other celestial bodies, shall be resolved by the States Parties to the Treaty either with the appropriate international organization or with one or more States members of that international organization, which are Parties to this Treaty.

Article XIV

1. This Treaty shall be open to all States for signature. Any State which does not sign this Treaty before its entry into force in accordance with paragraph 3 of this article may accede to it at any time.

2. This Treaty shall be subject to ratification by signatory States. Instruments of ratification and instruments of accession shall be deposited with the Governments of the Union of Soviet Socialist Republics, the United Kingdom of Great Britain and Northern Ireland and the United States of America, which are hereby designated the Depositary Governments.

3. This Treaty shall enter into force upon the deposit of instruments of ratification by five Governments including the Governments designated as Depositary Governments under this Treaty.

4. For States whose instruments of ratification or accession are deposited subsequent to the entry into force of this Treaty, it shall enter into force on the date of the deposit of their instruments of ratification or accession.

5. The Depositary Governments shall promptly inform all signatory and acceding States of the date of each signature, the date of deposit of each instrument of ratification of and accession to this Treaty, the date of its entry into force and other notices.

6. This Treaty shall be registered by the Depositary Governments pursuant to Article 102 of the Charter of the United Nations.

Article XV

Any State Party to the Treaty may propose amendments to this Treaty. Amendments shall enter into force for each State Party to the Treaty accepting the amendments upon their acceptance by a majority of the States Parties to the Treaty and thereafter for each remaining State Party to the Treaty on the date of acceptance by it.

Article XVI

Any State Party to the Treaty may give notice of its withdrawal from the Treaty one year after its entry into force by written notification to the Depositary Governments. Such withdrawal shall take effect one year from the date of receipt of this notification.

Article XVII

This Treaty, of which the Chinese, English, French, Russian and Spanish texts are equally authentic, shall be deposited in the archives of the Depositary Governments. Duly certified copies of this Treaty shall be transmitted by the Depositary Governments to the Governments of the signatory and acceding States.

IN WITNESS WHEREOF the undersigned, duly authorized, have signed this Treaty.

DONE in triplicate, at the cities of London, Moscow and Washington, D.C., the twenty-seventh day of January, one thousand nine hundred and sixty-seven."

Take careful note of the wording in Article VIII in which the United Nations, and the treaty's signatories, make it clear that the country that launches an object retains "jurisdiction", or ownership and responsibility, for that object. The treaty only identifies "space objects", not functioning and non-functioning spacecraft. There is no legally recognized international agreement that distinguishes between a "space object" that is functioning as a satellite and a piece of orbital debris that came originally from a functioning spacecraft or rocket. Who's to say that a particular debris object is junk and not a critical part of someone else's spacecraft that is merely dormant or not in use at the moment?

And Article VII, which defines who is responsible should a space object cause injury or damage, makes it more complicated still when today's economic globalization is considered. Consider the following hypothetical, yet realistic, scenario in which Country A contracts with a company in Country B to build a satellite that will be launched into space by Country C. Under which country shall the satellite be registered and therefore owned? The launching country? But what about the country that paid for it? Or the country in which it was manufactured?

Article VIII also makes it clear who is liable for the current crop of orbital debris. According to data provided to NASA by the US Space Surveillance Network in January 2012, Table A.1 shows which countries have "jurisdiction" (i.e. responsibility) for the largest of the debris objects in Earth orbit [2].

Table A.1. The countries or groups of countries that account for 90% of the large debris objects in Earth orbit today (China 22%, Russia 38%, USA 30%).

Country or organization	Payloads	Rocket bodies and known debris	Total
China	118	3,497	3,615
Commonwealth of Independent States (Russia, etc.)	1,417	4,670	6,087
European Space Agency	41	44	85
France	54	435	489
India	47	129	176
Japan	117	72	189
United States	1,158	3,692	4,850
Other	514	112	626
Total	3,466	12,651	16,117

If an individual country were to undertake a serious effort to reduce the threat of orbital debris by commissioning the removal of threatening debris objects, large or small, then would they be restricted to only remove those objects under their own jurisdiction? If so, then this would likely limit the remediation to only the large debris objects of known origin. Removing some of the half a million small objects might be prohibitive unless each piece's jurisdictional origin is well known and documented. In 1972, the United Nations passed the *Convention on International Liability for Damage Caused by Space Objects* in an attempt to clarify the international liabilities for space activities, but it unfortunately is still rather vague on many of the specifics that will have to be resolved before serious orbital debris remediation can begin. Below are excerpts from the UN agreement [1]:

"Article I (of the Convention on International Liability for Damage Caused by Space Objects)

For the purposes of this Convention:

(a) The term "damage" means loss of life, personal injury or other impairment of health; or loss of or damage to property of States or of persons, natural or juridical, or property of international intergovernmental organizations;

(b) The term "launching" includes attempted launching;

(c) The term "launching State" means:

(i) A State which launches or procures the launching of a space object;

(ii) A State from whose territory or facility a space object is launched;

(d) The term "space object" includes component parts of a space object as well as its launch vehicle and parts thereof.

Article II

A launching State shall be absolutely liable to pay compensation for damage caused by its space object on the surface of the earth or to aircraft flight.

Article III

In the event of damage being caused elsewhere than on the surface of the earth to a space object of one launching State or to persons or property on board such a space object by a space object of another launching State, the latter shall be liable only if the damage is due to its fault or the fault of persons for whom it is responsible.

Article IV

1. In the event of damage being caused elsewhere than on the surface of the earth to a space object of one launching State or to persons or property on board such a space object by a space object of another launching State, and of damage thereby being caused to a third State or to its natural or juridical persons, the first two

States shall be jointly and severally liable to the third State, to the extent indicated by the following:

(a) If the damage has been caused to the third State on the surface of the earth or to aircraft in flight, their liability to the third State shall be absolute;

(b) If the damage has been caused to a space object of the third State or to persons or property on board that space object elsewhere than on the surface of the earth, their liability to the third State shall be based on the fault of either of the first two States or on the fault of persons for whom either is responsible.

2. In all cases of joint and several liability referred to in paragraph 1 of this article, the burden of compensation for the damage shall be apportioned between the first two States in accordance with the extent to which they were at fault; if the extent of the fault of each of these States cannot be established, the burden of compensation shall be apportioned equally between them. Such apportionment shall be without prejudice to the right of the third State to seek the entire compensation due under this Convention from any or all of the launching States which are jointly and severally liable.

Article V

1. Whenever two or more States jointly launch a space object, they shall be jointly and severally liable for any damage caused.

2. A launching State which has paid compensation for damage shall have the right to present a claim for indemnification to other participants in the joint launching. The participants in a joint launching may conclude agreements regarding the apportioning among themselves of the financial obligation in respect of which they are jointly and severally liable. Such agreements shall be without prejudice to the right of a State sustaining damage to seek the entire compensation due under this Convention from any or all of the launching States which are jointly and severally liable.

3. A State from whose territory or facility a space object is launched shall be regarded as a participant in a joint launching.

Article VI

1. Subject to the provisions of paragraph 2 of this Article, exoneration from absolute liability shall be granted to the extent that a launching State establishes that the damage has resulted either wholly or partially from gross negligence or from an act or omission done with intent to cause damage on the part of a claimant State or of natural or juridical persons it represents.

2. No exoneration whatever shall be granted in cases where the damage has resulted from activities conducted by a launching

> State which are not in conformity with international law including, in particular, the Charter of the United Nations and the Treaty on Principles Governing the Activities of States in the Exploration and Use of Outer Space, including the Moon and Other Celestial Bodies."

To those who observed the effects of the Iridium/Cosmos collision (Chapter 1) in 2009, it was interesting to see virtually no outcry for compensation from either of the involved parties or any of the satellites now threatened by the large debris cloud created by the event. It is perhaps due to the wording of Article III that says that there must be negligence in order for there to be liability: ". . . the latter shall be liable only if the damage is due to its fault or the fault of persons for whom it is responsible."

And then there is the problem of unintentional consequences. If a country attempts to remove a debris object and fails, causing the object to enter a different orbit and inflict damage on a non-involved party's satellite, would it be considered an accident, a convenient test of a new anti-satellite weapon, or an act of war? Clearly, some new international agreement or legal regime is required before serious orbital debris clean-up can commence.

Legal Implications of Space Weaponization

The Outer Space Treaty makes it very clear that placing weapons of mass destruction in space is forbidden. Article IV specifically mentions that nuclear weapons are not allowed to be placed in Earth orbit or on the Moon. It says nothing about the development and testing of anti-satellite weapons such as those in the arsenals of Russia, China, and the United States. One need only recall another treaty, the 1928 Kellogg–Briand Pact, an international agreement in which signatory states promised not to use war to resolve "disputes or conflicts of whatever nature or of whatever origin they may be, which may arise among them", to become cynical about the prospects of any country abiding by these restrictions in time of war. The Kellogg–Briand Pact was signed by Germany, France, and the United States in 1928.

References

[1] *United Nations Treaties and Principles on Outer Space, Text of Treaties and Principles Governing the Activities of States in the Exploration and Use of Outer Space Adopted by the United Nations General Assembly.* United Nations Publication, ISBN 92-1-100900-6 (2002).
[2] *NASA Orbital Debris Quarterly News,* **16**(1) (2012).

Index